LIVESTOCK & MEATPACKING

Jeff Wilson

KALMBACH BOOKS

Kalmbach Books
21027 Crossroads Circle
Waukesha, Wisconsin 53186
www.Kalmbach.com/Books

Published in 2013
17 16 15 14 13   1 2 3 4 5

Manufactured in the United States of America

ISBN: 978-0-89024-867-6
EISBN: 978-0-89024-921-5

Editor: Randy Rehberg
Art Director: Tom Ford
Illustrator: Rick Johnson

Publisher's Cataloging-In-Publication Data

Wilson, Jeff, 1964-
   Livestock & meatpacking / Jeff Wilson.

      p. : ill. (some col.) ; cm. -- (Model railroader books) -- (Guide to
industries series)

   ISBN: 978-0-89024-867-6

   1. Railroads--United States--Livestock transportation--History. 2.
Livestock--Transportation--United States--History. 3. Packing-houses-
-United States--History. 4. Railroads--Models. I. Title. II. Title: Livestock
and meatpacking III. Series: Model railroader books.

HE2321.L7 W55 2012
338.17/6

CONTENTS

Introduction 4

Chapter One: **Industry history** 5

Chapter Two: **Stock cars** 10

Chapter Three: **Meat refrigerator cars** 23

Chapter Four: **Cattle pens and stockyards** 37

Chapter Five: **Packing plants and branch houses** 54

Chapter Six: **Livestock operations** 66

Chapter Seven: **Meat traffic** 78

Bibliography 86

Introduction

The livestock and meatpacking businesses, and their associated traffic—namely refrigerator cars of meat and meat products—were a key part of railroading from the 1800s through the 1960s. If you model any era during that time frame, you can replicate many facets of the industry regardless of the railroad or region you model.

A trio of Chicago & North Western F units hustle several loads of livestock eastward through Missouri Valley, Iowa, in 1947. *Henry J. McCord*

These industries provide many fascinating modeling possibilities. Cattle, hogs, and sheep traveled in stock cars. Dressed meat and finished meat products were carried to all corners of the country in a tremendous variety of refrigerator cars, many of which featured bright, colorful paint schemes. Structures included stock pens, stockyards, packing plants, and branch houses.

The meat business was no small player in American industry: by the 1930s, meatpacking ranked third behind only the auto and steel industries.

We'll take a look at the history of livestock production and the packing industry and see how railroads handled shipping both live animals and finished products.

Industry history

The histories of the livestock and meat processing industries are heavily intertwined with the development of railroads in the United States. As the packing industry grew in the late 1800s, railroads became a vital link in both getting animals from rangelands to packing plants and delivering dressed meat and meat products from packing plants to consumers throughout the country.

Today, the word *livestock* to many people means beef cattle. However, other types of animals were transported in large numbers, namely pigs and sheep but horses and mules as well.

A New York Central livestock special leans into a long curve by Newark, N.Y., in the 1940s. *S.K. Bolton Jr.*

5

This late 1800s illustration depicts cattlemen on horseback meeting buyers arriving by train at Ellsworth, Kan., where cattle drives met the Kansas Pacific (later Union Pacific). *Union Pacific*

By the mid-1800s, the meatpacking business was already a significant industry in the United States. Most large cities had several packing plants where animals were butchered and processed. With no practical means of refrigeration, most meat could not be sold fresh or raw—it had to be cured or preserved. This was usually done by salting or smoking.

Various types of meat could be cured this way, but cured pork was preferred for taste over preserved beef, and pork dominated the industry: by 1870, pork accounted for 91 percent of packing house output in the United States, compared to just 3 percent for beef. At the time, New York City led the country in beef production simply because it was the largest metro area: most was sold and consumed locally.

In the early 1800s, Cincinnati had established itself as the country's prominent producer of pork, earning the not-always-complimentary nickname of *Porkopolis*. The rise of cattle and beef production would come later, thanks in part to the spread of railroads and the coming of refrigeration.

Late 1800s
By the post-Civil War era, railroads were expanding at a rapid rate. The first transcontinental railroad was completed in 1869, and many other lines were stretching westward across the Mississippi. Thousands of miles of branch lines were being built in the Midwest and granger states in an effort to tap into farm markets and bring settlers and goods to new towns and cities across the West.

In that era, most meat processing was still done locally. Refrigeration wasn't available, so meat had to either be consumed quickly or preserved. For railroads, this meant that transportation of the animals themselves—not the finished products—was the priority.

Southwestern and western rangelands weren't good for growing crops, but the wide-open spaces were ideal for grazing large herds of cattle. Getting those cattle to eastern markets became the job of the railroads. A general pattern was soon established that would last through the 20th century, in which most cattle were raised west of the Mississippi River but most food was consumed east of the Mississippi. Railroad traffic flow of livestock and meat followed this pattern.

Before western railroads were fully established, cattle drives were the only way to move livestock on their way eastward. Between 1867 and 1872, more than 3 million head of longhorn cattle were driven from Texas along the Chisholm Trail to Abilene, Kan. The cattle were shipped on the Kansas Pacific Railroad (later part of the Union Pacific) to points east. Long-distance cattle drives continued to Abilene and Dodge City, Kan., on the Santa Fe Trail until 1885, when the rail network had expanded far enough to make drives impractical, with all-rail shipment becoming possible from all grazing areas.

Centralized packing
By the 1860s, Chicago had emerged as a dominant rail center. The city was the country's most important traffic gateway between established eastern markets and the expanding West.

Chicago's Union Stock Yards, which would eventually become the largest in the country, opened on Christmas Day in 1865. Formed by a group of nine railroads, the stockyards provided a centralized location for selling, buying, and processing livestock. Farmers and western ranchers would send animals to the stockyard, where representatives of eastern packers would purchase them.

Chicago Union processed 1.5 million head of livestock in its first year of operation, and by the 1870s, it was handling more than 2 million head per year. Most of the arriving animals were passing through on their way to eastern markets, with only about 250,000 cattle per year slaughtered in the city—most for local consumption.

The coming of economical mechanical refrigeration in the 1870s and 1880s revolutionized the meatpacking industry. Refrigeration allowed packing plants to expand to large scale operations beyond local sales. Coupled with the development and improvement of the ice-bunker refrigerator car (see Chapter 3), packing companies could now ship their products throughout the country.

Packing companies began to establish plants next to union

stockyards, which provided a ready supply of livestock. Chicago became the country's preeminent meat-producing center. Gustavus Swift opened his Chicago packing plant in 1875, the first of several major packing companies to do so. Armour and Nelson Morris followed in 1882, with Cudahy and Wilson following later in the decade.

For railroads, the 1880s saw a shift in traffic from stock cars to refrigerator cars of dressed meat. As an example, from 1882 to 1886, shipments of live cattle to New York City dropped from 366,500 to 280,200 carloads, while shipments of dressed beef in reefers jumped from 2,600 to 69,800.

Meatpacking then shifted farther west. By 1900, the top five meatpacking states were Illinois, Kansas, Nebraska, Indiana, and Missouri. Only Illinois had made that list 40 years earlier.

Railroads, having already invested heavily in stock cars, initially resisted this change. Railroads were reluctant to invest in refrigerator cars, as they were more expensive to build and maintain than stock cars. Also, since a dressed carcass weighed significantly less than a live animal, railroads had less to ship. This led to private ownership of most refrigerator cars, as described in Chapter 3.

By 1900, Chicago packing companies were slaughtering more than 2.2 million animals annually (accounting for about 80 percent of United States meat consumption), with 14 million animals passing through the city's union stockyards. The stockyards covered 475 acres with a pen capacity of 75,000 cattle, 50,000 sheep, 300,000 hogs, and 5,000 horses. Railroads, with about 68,000 refrigerator cars in service, carried the products of these packers throughout the country.

Refrigeration brought another change to the industry, as production shifted from hogs to beef

A Sioux City Terminal Railway steam switcher pulls a cut of stock cars into position at the city's union stockyards in the early 1900s. *Courtesy Sioux City Historical Society*

cattle, reflecting the preference of many for steak over pork. By the 1890s, beef had grown to 34 percent of the market, compared to 42 percent for pork, a trend that would continue into the 1900s.

Growth of the Big Five

In the early 20th century, meatpacking had become the second-largest industry in the country, trailing only the steel industry. Five companies dominated: Swift, Armour, Nelson Morris, Cudahy, and Wilson (Schwarzchild & Sulzberger until 1917). Known as the Big Five (the Big Four after Armour purchased Nelson Morris in 1923), these companies controlled two-thirds of the country's fresh beef market.

The Federal Trade Commission found that these companies were illegally conspiring to corner the market, and by 1920, the packers had to divest themselves from non-meat-production ventures including stockyard and retail ownership, as well as refrigerator car leasing, reducing Armour's dominance in that field.

In the early 1900s, Chicago's union stockyard was the largest in the country, and the output of its packing companies also was tops

in the country. However, Chicago was by no means the only major meatpacking center. Large union stockyards operated in several other cities (see Chapter 4).

All the Big Four companies had multiple packing plants around the country, most located near large union stockyards. In addition, many small and regional packing companies had significant operations.

Traffic patterns

Through the mid-1900s, the country's pattern of livestock production and meat distribution was well established. Cattle were raised on western ranches as well as farms throughout the country. When an owner was ready to sell, animals were shipped by rail or truck to a nearby union stockyard, where they were purchased by either a packing company or by a rancher, feedlot, or another farmer for additional fattening.

Long-distance hauls of livestock were common. Railroads maintained stock pens all along their routes to enable the loading of stock cars, and they also operated large facilities for trainloads of animals to be let off, watered, and fed to comply with handling

Packers sometimes marked special shipments with banners on refrigerator cars. This wood Cudahy reefer, parked outside the company's plant in North Salt Lake, Utah, wears the classic Old Dutch Cleanser billboard scheme. *Courtesy Utah State Historical Society*

legislation, mainly the 28-hour law that required periodic rest of animals in transit.

Many railroads operated dedicated livestock trains, and special movements of animals were common, especially in the fall and spring as animals were moved between summer and winter grazing lands.

Large packing companies, mainly located near large union stockyards, butchered the animals and prepped them by dividing them into halves or quarters (known as dressed meat). Packers would also process some meat into finished products or ship the dressed meat to regional branch houses for further processing or sale.

Railroads handled much of this traffic, transporting refrigerator cars from packing plants to regional branch houses. Regular shipments included dedicated trains of meat reefers operating out of major packing centers.

The phrase "everything but the squeal" described how all parts of each animal were used. By-products were processed or rendered to produce lard, oleo-margarine, fertilizer, pet food, soap, pharmaceuticals, and many other products. Hides were shipped to tanneries to make leather products. Depending upon the size of the packing plant, these products were made on site or the by-products were shipped to rendering plants, often by rail.

Changing market

As roads and highways improved in the 1930s and '40, livestock and meat traffic began shifting from railroads to trucks. This came first for shorter hauls and then—with the coming of interstate high-ways—for long hauls as well.

The 1950s saw the growth of feedlots in the West, Midwest, and Southwest. At feedlots, a large number of animals would be kept in a very small area and fed grain, compared to the traditional herds of animals grazing on the range or grasslands. Feedlots hadn't been practical until this time: keeping large numbers of animals in a small area wasn't feasible because of the high risk of disease, which could spread quickly among confined animals. The use of penicillin and other antibiotics minimized disease risks.

Coupled with low grain prices in the 1950s, the period saw dramatic growth in the number of cattle on farms, but a drop in the overall number of farms. By the early 1960s, two-thirds of beef cattle in the country were being raised on grain.

Before World War II, around 90 percent of livestock was sold through union stockyards. By 1950, this had dropped to 75 percent, and by 1960, it was less than 50 percent, with more farmers and feedlot operators selling directly to packers.

Chicago Union Stock Yards, which handled more than 10 million head a year into the 1930s, saw a heavy decline in numbers through the 1940s, as stockyards in other cities took away business. The numbers in Chicago remained steady, above 5 million annually into the early 1950s, but then began dropping significantly. In 1955, Omaha took over from Chicago as the country's largest union stockyard, with 6.7 million head, but the numbers there would also drop through the 1960s.

The aging plants of most large packing companies were growing obsolete, and packers were beginning to see that it made economic sense to relocate new plants closer to feedlots and other sources of livestock.

Cudahy, in 1954, was the first major packer to close its Chicago plant. Swift followed in 1958 and Armour in 1959, and all of Chicago's once-mighty meat packers shut their doors by 1960.

The traditional market process of packing companies, butchers, and meat markets had begun to change with products marketed directly to supermarkets and retailers. This started in 1926 when Hormel introduced the first canned ham, a product soon followed by many others, and with Spam, a canned pork prod-uct that debuted in 1937.

This 1947 view of Sioux City, Iowa, shows the Cudahy and Armour packing plants in the distance. The city's sprawling union stockyard at left was still doing a brisk business, and the sea of stock cars shows that railroads were still handling plenty of livestock traffic. *Henry J. McCord*

Other innovators included Oscar Mayer, which offered the first packaged, sliced bacon in 1924, and also introduced the first vacuum-packed sliced cold cuts in the late 1940s. These products foreshadowed evolution in processing meat by providing finished products ready for retail.

The company most responsible for the industry's new era was Iowa Beef Packers (IBP), which opened its first plant in Denison, Iowa, in 1961. The company changed the market in three major ways. First, it built plants in rural areas and small towns next to feedlots. This cut transport expenses as no long-distance movement of livestock was needed. Buying cattle directly avoided union stockyards and saved commission fees.

Secondly, IBP divided jobs into additional steps among more levels, giving many workers a single task, so that untrained (and thus cheaper) labor could fill most jobs. This, combined with using non-union workers, saved labor costs.

Finally, IBP processed much of its meat to the final cut and then packaged it ready for retail markets, without having to go through a branch house (or even a local butcher). This enabled IBP to sell directly to retailers, earning money that would have gone to wholesalers or other middlemen.

The era of livestock on rails was drawing to a close in 1977, when these triple-deck hog cars of Union Pacific's California Livestock Special rolled west through Dry Lake, Nev. *Steve Patterson*

By the late 1960s, IBP (which became Iowa Beef Processors in 1970) operated eight packing plants in the Midwest. Despite major labor problems, including some violent strikes, IBP grew to become the nation's largest beef producer and No. 2 pork producer.

Other smaller companies followed IBP's lead, which cut into the sales of the major packers. By 1970, only 16 percent of livestock sold in the United States passed through a union stockyard.

End of rail operations

During this time, rail carloadings for livestock and meat products were falling dramatically. Other than a few dedicated markets, the shipping of livestock by rail was mostly gone by 1970.

The downturn of the traditional traffic from packing plant to branch house had eliminated a lot of meat reefer traffic by the 1960s. Ice-bunker reeferss were being phased out—for fruits and vegetables as well as meat—through the 1960s. Shippers and railroads were not willing to invest in expensive mechanical reefers for a dwindling market, and semitrailers took over what little rail traffic remained by that time.

The most telling sign came on August 1, 1971, when Chicago Union Stock Yards ceased operations. The closing of what was once the world's largest stockyard symbolized the dramatic changes in the industry and represented the end of a colorful, busy part of railroading.

Stock cars

A Burlington Route local spots a pair of stock cars at a Wyoming stock pen in 1955. The 40-foot cars, part of CB&Q class SM-19B, were typical of modern 1950s cars, with steel underframes, ends, and roofs. The single-deck cars were built in 1949 at the railroad's Havelock shops. They have a 9'-0" interior height and doors 5'-0" wide. *William A. Akin*

For more than 100 years, stock cars were a common way for livestock to travel from farm and ranch to market. Blocks of stock cars could be found in many freight trains, and solid trains of stock cars were common during the fall and spring, especially in the Midwest and West. They were distinctive for their slatted sides as well as their sound and—depending which way the wind was blowing—their smell.

TWO

The Hicks car was one of the early (late 1800s) patent humane stock cars, with built-in feed and water troughs. The car featured all-wood construction with a truss-rod underframe and archbar trucks. *David P. Morgan Library collection*

Early stock cars

Railroads have historically preferred general-purpose rolling stock whenever possible. The ubiquitous boxcar was the mainstay of the freight fleet from the mid-1800s through the late 1900s because it could be used for hauling many different products, helping ensure that it would spend more time traveling while loaded than empty. Railroads have always been reluctant to build specialty cars for single or limited commodities. This was certainly the case with stock cars and livestock, since such cars generally operated while loaded only half the time. In addition, stock traffic tended to be seasonal, with more traffic in the fall and spring, so stock cars could sit unused for long periods of time.

As noted in Chapter 1, railroads began moving livestock by rail in the 1830s, shortly after rail lines were first established in the United States. Most of these were relatively short-haul, low-speed operations. At the time, common boxcars were often used for shipping livestock. As can be imagined, even a short trip or two

This Milwaukee Road double-deck car is typical of stock cars at the turn of the 20th century. The car has all-wood construction, roof hatches, end doors, and archbar trucks. Note the flatcar-style construction with the car posts in pockets on the side sill. *David P. Morgan Library collection*

with live animals, with the cars' wood floors and walls, made these cars undesirable for other lading.

As rail lines expanded before and after the Civil War, a major problem in hauling live animals was train speed. The slow speeds of the era meant that even a relatively short trip by today's standards could take dozens of hours.

By the post-Civil War era, as the transcontinental railroad was completed and other railroads expanded across the Mississippi River to the west, railroads began building cars designed specifically for carrying live animals.

A number of designs were attempted, but by the 1880s, railroads had settled on the basic stock car design that would, with

Mather stock cars could be found on railroads across the country. They are easily spotted by their U-channel diagonals, Z-shaped posts, and a pair of diagonal straps in the panels closest to the car ends. This 40-foot car is awaiting loading in Texas in 1964. *Donald Sims*

The Santa Fe operated more stock cars than any other railroad. The Illinois Railway Museum refurbished ATSF No. 60394, one of 500 class Sk-T single-deck cars built by Pennsylvania Car Co. in 1929. *Jeff Wilson*

around 1880 took about five days). Weak animals were often gored, trampled to death, or died of other causes en route, and even healthy animals lost excessive body weight (*shrinkage* in industry terms) en route.

The eventual solution was the 28-hour law (see Chapter 6), which required railroads to periodically off-load livestock for rest and feeding while in transit. However, this law didn't truly take effect until 1906. Until that time, the best attempt to keep animals healthy and alive was the advent of the so-called "humane" (or palace) stock car. These were attempts at building cars with built-in feed and watering troughs. Humane cars followed several designs and patents, and although several thousand were placed in service, they met with only moderate success, as feed and water often were never refilled as intended.

Following the advent of the 28-hour law, builders abandoned

refinements, be used through the end of the rail/livestock era. Cars of the era featured all-wood construction. Stock car bodies had horizontal wood slatted sides mounted on side trusses of various designs. The ends and

roofs were also wood.

During that period, as cattle were being transported longer and longer distances, animals were confined in cars for many days, in some cases with no relief. (A Chicago to New York trip

Major stock car fleets
Railroads with at least 1,000 stock cars in 1943

RR	1932	1943	1950	1955	1962	1966	1971	1977	1981	1986
ATSF	9,318	7,906	7,295	7,709	6,370	4,713	2,317	0	0	0
B&O	1,691	1,198	1,195	900	1,021	368	10	0	0	—
BN	—	—	—	—	—	—	3,104	2,197	332	0
CN	4,712	2,994	2,986	2,873	2,249	1,142	1,485	914	473	62
CP	3,654	2,721	3,357	2,814	2,111	1,620	1,297	1,067	875	601
CNW[1]	5,230	4,630	3,852	2,059	1,015	737	55	0	0	0
CBQ	5,926	3,463	3,573	3,493	3,235	1,968	—[2]	—	—	—
MILW	4,301	3,901	3,721	3,315	3,061	1,846	811	34	0	0
CRIP	3,271	1,528	1,238	838	629	422	64	0	—	—
DRGW	1,364	1,133	1,262	909	791	778	416	3	0	0
GN	3,468	1,948	2,045	1,718	1,981	1,895	—[2]	—	—	—
IC	1,749	1,158	1,106	712	219	202	85	0	0	0
MoPac	2,864	1,399	1,528	1,517	792	432	145	0	0	0
NYC	5,427	1,366	1,687	1,600	596	45	—[3]	—	—	—
NP	2,770	1,814	1,717	1,724	1,618	1,522	—[2]	—	—	—
PRR	3,319	2,438	2,356	1,299	353	411	—[3]	—	—	—
PC	—	—	—	—	—	—	110	—	—	—
SP[4]	7,037	4,662	3,072	2,302	761	92	0	0	0	0
UP	6,347	5,972	4,486	3,336	3,249	2,545	2,102	1,024	502	235
Private car owners										
ASEX	200	200	200	200	51	0	—	—	—	—
Mather[5]	100	958	21	390	—	—	—	—	—	—
SLSX	no data	863	716	494	193	51	—	—	—	—
QLSX	no data	497	no data	483	—	—	—	—	—	—

[1] Includes Chicago, St. Paul, Minneapolis & Omaha
[2] To Burlington Northern
[3] To Penn Central
[4] Includes Texas & New Orleans and other subsidiaries
[5] Mather cars were often leased to railroads and wore a railroad's reporting marks

ASEX — Armour Stock Express
SLSX — Swift Livestock Express
QSLX — Quaker City Livestock Express (also includes GASX and General American)

the built-in troughs and went back to basic construction. Many early stock cars were built atop existing flatcars or followed the same design, with the body superstructure built atop a flat bed. The vertical side members on these cars fit into the stake pockets along the flatcar side sills. The Milwaukee Road stock car pictured on page 11 is an example.

These cars were used to carry all types of livestock, including cattle, sheep, hogs, mules, and horses. High-value animals (race horses and breeding horses) were sometimes shipped in horse cars that resembled baggage cars.

Construction
As with other "house" (enclosed) cars of the late 1800s, stock cars had wood floors and underframes, with steel truss rods providing strength to the underbody. Stock cars generally followed boxcar developments over time, but—being single-commodity cars—stock cars often used simpler, cheaper construction techniques and materials. Because their seasonal use left them idle for long periods, railroads would

Stock cars in service by year
1880:	28,000
1890:	57,000
1910:	79,000
1932:	97,000
1941:	54,000
1960:	31,000

Source: *Official Railway Equipment Register*, various issues

rebuild and maintain older stock cars, giving many of them long service lives compared to other contemporary cars.

All-wood construction was common into the 1910s. Wood remained the common material

The narrow gauge Denver & Rio Grande Western operated a sizeable fleet of stock cars to move cattle to various rangelands along its route. Number 5670, a 30-foot, 25-ton car, is at Durango, Colo., in 1949. *John Allen*

for roofs and ends, while underframes gradually changed to steel.

Typical stock cars were 28 feet long into the 1880s. The length increased to 32 and then 36 feet shortly after the turn of the 20th century. Although boxcars were commonly 40 feet long by 1920, many stock cars continued to be built to 36 feet. It wasn't until the late 1940s when the number of 40-foot stock cars in service began to equal to the number of 36-footers.

Stock car height varied greatly among railroads, and, except for some double-deck cars, they were often shorter in height than contemporary boxcars.

By the 1950s, declining stock loadings meant few new stock cars were being built. Instead, most railroads rebuilt existing cars, or—if needed—rebuilt stock cars from old boxcars.

Secondary uses
Although stock cars are single-commodity cars, railroads were always looking for ways to improve their utilization by

finding materials they could haul on return trips or in the off-season when they would be sitting idle.

Hay bales and railroad ties were two common stock-car loads. Lumber was also sometimes carried. In busy grain seasons, stock cars would sometimes be lined with plywood and used to haul grain in a similar manner as boxcars.

Coke was another commodity hauled in stock cars. In fact, some early stock cars were outfitted with drop-bottom doors (similar to drop-bottom gondolas) to make it easier to unload the coke. Examples were Santa Fe's class Sk-H, -K, -N, and -P cars built from 1906 through 1924. These cars, which were used this way into the 1930s, also had loading hatches in their wood roofs to enable easier loading of coke.

Double-deck cars
To increase a car's capacity for hauling calves, sheep, and hogs, many cars were equipped with a second deck. This effectively doubled the car's capacity without

increasing the car's size, and even with a load of animals on the second deck, it wouldn't approach the car's weight limit.

Permanent double-deck cars were built starting in the 1800s, and convertible double-deck cars began appearing in the early 1900s. This variation featured a floor that could be raised up to the roof when not needed and lowered into position when required. Most convertible cars had a crank mechanism on the outside of one side for doing this.

You can often spot a double-deck car by the doors: many double-deck cars have a pair of doors (and door tracks), one above the other. You can also usually see the floor through the side, with the second-deck floor interrupting the slat pattern in the sides.

Stock car fleets
Railroads generally built their fleets of stock cars following the amount of livestock they carried. The table on page 13 shows the number of stock cars various

Double-deck Chicago, Burlington & Quincy No. 58214 carries a load of sheep in the early 1960s. The 36-foot car has wood ends and a steel underframe. Built in 1924, it was one of 1,500 class SM-16 cars on the railroad. *Big Four Graphics*

railroads had in service over the years. The largest fleets were owned by western railroads that ran through ranch and farm territory and served the major packing and stockyard centers, with Santa Fe, Union Pacific, Southern Pacific, Chicago & North Western, and Chicago, Burlington & Quincy leading the lists.

In the East, the Pennsylvania Railroad and New York Central had the largest stock car fleets, largely because they were the railroads that carried the most livestock from Chicago and western connections to New York and other eastern markets.

Many private owners also operated stock cars, most notably Mather, which operated cars and also leased cars to railroads and packing companies. Among packing companies, both Armour and Swift operated small fleets of stock cars. As it did with its reefers, Swift sold its stock car fleet in 1930 to General American, which leased the cars back to Swift. These cars bore SLSX reporting marks, and many

Northern Pacific had a series of distinctive 43-foot stock cars with radial roofs and sign boards at the top of the sides. Number 83457, here on the Burlington Northern at Minot, N.D., in 1974, was built by Ryan Car Co. in 1930. Cars of the same design were also built by the railroad's Laurel and Como shops. *Jim Hediger*

carried Swift Livestock Express lettering.

Railroads often built stock cars to their own design and also built them from older boxcars. This meant each railroad's stock cars had a distinct appearance. Spotting features include car height

and length, type of end and roof, side truss pattern, and door style and width.

Stock cars were built by major builders such as American Car & Foundry, Pullman-Standard, General American, and Pressed Steel Car Co., as well as many

Number 27200 is one of 200 stock cars the New York Central rebuilt from old 40-foot USRA single-sheathed boxcars in 1948. The double-deck cars retained their ends and side trusses from the USRA car and are similar to 500 convertible double-deck cars rebuilt a year earlier. *New York Central*

This 40-foot Sioux City Terminal Ry. car was built by General American in 1931. Straw bedding is visible at the door and through the bottom slats. *David P. Morgan Library collection*

smaller builders, including Ralston, Ryan Car Co., Pennsylvania Car Co., and others.

Because so many stock cars were rebuilt from older boxcars and because few stock cars were built when compared to boxcars, standard car designs weren't followed in large numbers for stock cars.

In 1927, the American Railway Association adopted a recommended car design. The design called for a 40-foot, 40- or 50-ton car with wood roof and ends. This AAR (American Association of Railroads) design was upgraded in 1951 and featured a 40-foot car with Dreadnaught ends and a diagonal-panel steel roof. The Union Pacific S-40-15 is an example that follows this design.

Through the 1920s and even later, stock car roofs were often wood, with tongue-and-groove construction for a tight fit. These roofs were sometimes covered with tarpaper, canvas, or thin metal sheets. This was less expensive than the fabricated steel roofs of contemporary boxcars, and leaks weren't as critical with stock cars as with boxcars.

Wood ends were common on stock cars during that period, but cars rebuilt from older steel-end boxcars would generally retain those ends. Some cars had small or large end doors (called *drovers doors*), which allowed a handler (drover) access to the car without having to open a large side door.

A car's side forms a truss, which is its most distinctive (and noticeable) feature. Look for the number of panels on each side, the pattern of the diagonals (which way they face on each

Mather built several 50-foot stock cars by combining two shorter cars. Number 50237 is one of 40 such cars leased by the Chicago, Burlington & Quincy in the 1960s. *Virl Davis, Hol Wagner collection*

end), the arrangement of additional bracing on the end panels, and the materials used (some later cars used steel for slats). The width of the horizontal boards and slatted openings also varied.

Some railroads added lettering boards over the truss members. The boards could feature just the railroad's name and/or logo, or they could include the capacity, reporting marks, and dimensional data as well.

Trucks and brakes largely followed the style of other contemporary cars, with older K brakes being upgraded to AB brakes, and old truck designs such as Andrews upgraded to AAR cast trucks.

Mather stock cars

A number of early cars were patented by various manufacturers, including Hicks, Burton, and Canda. However, Mather was the manufacturer most associated with stock car design and operation through the steam era.

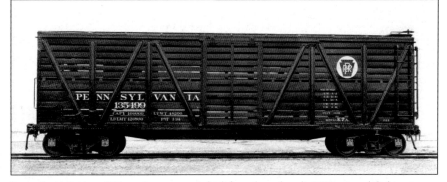

Pennsylvania's class K7A stock cars were rebuilt from the railroad's X24 automobile boxcars, which by the 1930s were too small for that service. This car was rebuilt in 1934, and more than 1,400 were in service through the 1940s. *Pennsylvania Railroad*

Alonzo C. Mather was a businessman who, by a chance encounter with a livestock train in the early 1880s, decided to design and build his own version of a stock car. By the mid-1890s, Mather put more than 3,000 stock cars in service.

Early Mather cars, like other patent cars, had built-in feed and water troughs, but these were abandoned by the turn of the 20th century as railroads and shippers turned to rest stops and hours-in-transit rules to solve the problems of shrinkage and in-transit deaths of cattle.

Mather's method, which he later also applied to boxcars and refrigerator cars, was to use basic, sturdy designs and inexpensive building materials—standard lumber and rolled-steel components instead of specialty (cast or fabricated) hardware and materials—and to lease cars

In the late 1930s, the Pennsylvania Railroad rebuilt more than 100 old 50-foot round-roof automobile boxcars into class K9A stock cars. They were distinctive, with rectangular cutouts, slats, and screen-wire covering. *Chuck Blardone*

Number 14266 is from the first batch of 400 stock cars (of 900) that the Chicago & North Western rebuilt from old single-sheathed boxcars in 1954 and 1955. Note the temporary upper deck (boards sticking through slats) added to carry bales of hay. *Lloyd Keyser*

as needed to railroads and shippers.

The Mather car fleet grew through the 1920s, and even in the Depression, railroads leased cars from Mather (keeping the company's Chicago production plant going), as doing so was more affordable than buying cars or building their own. Mather built 36-foot stock cars into the 1930s and then switched to a 40-foot design.

The stock cars were wood but rode on steel center sills made from a pair of I beams. Mather cars are identifiable by their sides, which had three panels on each side of the door, with U-shaped steel channels for diagonals and vertical Z-shaped posts. These angles and posts extended to the bottom of the side sill.

Car ends were wood with horizontal planks mounted on vertical members. Many cars had open slats at the tops of each end, while other cars had solid ends.

Through the 1930s, most Mather cars rode on Andrews trucks, but some were built or refitted with AAR cast sideframes.

Mather cars on long-term leases were generally painted in the leasing railroad's paint scheme with the railroad's reporting marks. Cars on short-term leases carried Mather's own MSCX reporting marks (NISX after Mather's acquisition by North American Car Co.). Because cars were constantly returning from or going out on lease, it is difficult to track the exact number of cars in service for Mather and various lessors at any given time.

Mather's 36-foot cars were common through the steam era, and many 40-foot cars lasted through Mather's acquisition by North American in the 1960s to the end of regular cattle shipments.

In the late 1950s, Mather rebuilt some of its 40-foot cars into 50-footers. These cars had two sets of doors, and although a

The Union Pacific's S-40-16 stock cars, built in 1964, had three decks for maximizing space for hauling hogs. They were more than a foot taller than most earlier stock cars. *Union Pacific*

Car capacity

The number of animals that could be safely loaded in a stock car varied greatly, depending upon the size and weight of the animals and the length of the car. This chart summarizes a table that appeared in a livestock loading and handling booklet published by the Santa Fe in the mid-1940s. Other railroads had similar guides.

Cattle						
Average weight per head	300	500	700	1,000	1,200	1,400
36-foot car, single deck	60	42	33	25	22	19
40-foot car, single deck	67	46	37	27	23	21
Hogs						
Average weight per head	100	150	200	250	300	400
36-foot car, single deck	130	100	79	68	60	48
36-foot car, double deck	250	190	150	130	112	88
40-foot car, single deck	145	110	88	76	66	54
40-foot car, double deck	380	210	167	144	124	100
Horses						
Average weight per head	800	1,000	1,200	1,400	1,600	1,800
36-foot car, single deck	27	24	22	19	18	17
40-foot car, single deck	30	27	24	21	19	18
Sheep or lambs						
Average weight per head	50	75	100	125	150	180
36-foot car, single deck	150	125	105	95	85	75
36-foot car, double deck	300	250	210	190	170	150
40-foot car, single deck	165	138	116	105	94	82
40-foot car, double deck	330	276	236	210	188	164

Poultry cars

Poultry cars were a small part of the stock car market, much less the overall freight car fleet. Just over 2,800 poultry cars were in service in 1932 (only 27 of which were railroad owned), amounting to just under 3 percent of all stock cars in service.

Cars designed specifically for chickens, turkeys, and other birds began appearing in the 1880s. Into the 1900s, most cars were operated by Live Poultry Transit Co. (earlier, the Live Poultry Transportation Co.). Another company, the Palace Poultry Car Co., began operations in 1924 and was shortly acquired by North American Car Co. This became the dominant owner of poultry cars, and North American eventually acquired LPTC in 1930.

North American spun off its poultry cars to the Poultry Transit Co. in 1944, and business slowly declined until that company ceased operations in 1956.

Cars were built by several companies, including Haskell & Barker (later part of Pullman), Pullman, and American Car & Foundry.

Poultry cars were designed with a series of racks to hold cages. The outsides were mesh screen allowing for ventilation. The center of each car housed an 8-foot-wide room for an attendant, who was responsible for watering and feeding the birds in transit. The room included a bunk, stove, and sink.

A tank atop the car held water; feed came from a bin below the attendant's room floor.

Poultry cars were largely used in the East, carrying birds to market in large cities.

The destination for many poultry cars on the East Coast was the West 60th Street poultry terminal in New York City. *New York Central*

Poultry cars had screened sides and shelves for holding crates of chickens. They carried feed and water, and an attendant rode in the center compartment. This car was built by American Car & Foundry in 1933. *American Car & Foundry*

few were single-deck cars, most were double-decked.

High-speed cars

Increased train speed was a factor in avoiding rest stops. As Chapter 7 explains, many railroads operated stock trains on expedited schedules. The Union Pacific did this with its Daylight Livestock (DLS) train, inaugurated in 1947, which ran from Salt Lake City to Los Angeles. The Union Pacific upgraded a group of its stock cars specifically for high-speed service: 300 initially, followed by another 500 cars.

Cars for this service came from the railroad's fleet of S-40-10 and S-40-12 cars. Cars for DLS service featured slats on the inside that were perfectly smooth, 6-foot-wide doors, upgraded air brakes, cushioned draft gear, and Timken roller-bearing trucks with bolster snubbers and high-speed steel wheels.

The cars also received a new paint scheme, setting them apart from the UP's standard mineral-red scheme. The roofs and ends were painted aluminum in an effort to reflect heat and keep interiors cooler. Car sides were yellow with red lettering and had an additional Livestock Dispatch sign board below the railroad name.

A later example of high-speed cars was the Union Pacific triple-deck S-40-16 car built in 1964. This car had a 10'-8" interior height that allowed a third deck for loading hogs. (Earlier cars had interior heights from 8'-10" to 9'-4".)

The 270 cars in this group were in dedicated service for hauling hogs on the DLS to Clougherty Packing Co. in Vernon (Los Angeles), Calif. Like the earlier cars, these had roller-bearing trucks and wore the aluminum and yellow paint scheme.

In the early 1970s, the UP upgraded a batch of triple-deck cars with in-car water and feed troughs along with adjustable slats. By furnishing food and water in the cars, the hogs were exempt from 28- or 36-hour restrictions.

Pig Palace cars

The Northern Pacific built 200 double-deck stock cars in 1958 and dubbed them Pig Palace cars. The cars, rebuilt from older boxcars, were all steel except for the floors. The sides featured a double set of steel slats with moveable outer slats. In cold weather, the outer slats could be moved up to cover the openings, which provided more temperature control compared to a standard stock car.

Like the UP cars, the NP's Pig Palace cars rode on roller-bearing trucks with snubbers and had rubber draft gear for a smoother ride at high speeds.

These led to the production in 1966 of what turned out to be among the last new stock cars built, NP's 86-foot Big Pig Palace cars. A prototype car was built in 1964 by combining two 40-foot Pig Palace cars to make an 85-foot car. The design proved successful, leading to the new cars, which were built by Ortner. The cars looked much like the new high-cube boxcars of the day, but the stock cars were standard height.

Ortner built 22 Big Pig Palace cars for NP, and similar cars were operated by Cattle Car Leasing Co. (16 cars), General American (40 cars), and Rochester Independent Packing Co. (5 cars).

The cars had all-steel side construction with welded joints to ensure smooth interior surfaces for the safety of the hogs. Cars were double-decked, and each deck was divided into two compartments with doors on each side for each compartment. The lower slatted openings on each deck were covered with screen to help retain bedding. The NP cars survived well into the Burlington Northern era, lasting into the 1980s.

The Northern Pacific's Pig Palace cars, rebuilt from older boxcars in 1958, were all steel (except for flooring), had adjustable slats, and rode on roller-bearing trucks. *Northern Pacific*

Among the last all-new stock cars built were the Northern Pacific's 86-foot Big Pig Palace cars. They were built by Ortner in 1966. *Ortner Freight Car*

The cars were intended for westbound shipments of hogs from South St. Paul and other points to West Coast markets. To get better car utilization, the cars could carry sheep or calves eastbound.

Southern Pacific also had a pair of 86-foot, double-deck cars built in 1964. Called Stock Palace cars by the SP, they differed from the NP cars by having doors at the end of each side (instead of inset), and they were excess-height cars, standing 17 feet tall. They were also equipped with Hydra-Cushion underframes.

HOGX cars

The last cars built for domestic livestock traffic were for carrying

In the 1970s, Union Pacific rebuilt several triple-deck cars with adjustable slats (controlled by levers at bottom of panels) and in-car water and feed troughs. *Jim Hediger*

hogs on the Union Pacific from North Platte, Neb., to Los Angeles (to Clougherty), on what UP termed the *hog train* or *ham train*. Gunderson rebuilt 90 60-foot boxcars into stock cars for the service. The cars rode the head end of intermodal train NPLAZ, pausing to water the hogs at Dry Lake, Nev.

The triple-deck cars had steel-slatted sides and a loading door at one end of each side. The service lasted into 1994, and when the service ended, it closed the book on transporting livestock by rail.

Modeling

The good news is that there are several good stock car models on the market. The bad news is that, because there was so much variation among prototype cars, the available models only represent a small portion of the real cars in operation.

In HO scale, good plastic models include the Proto 2000 Mather 40-foot car, which is accurate for several prototypes. The Accurail 40-foot car is based on a Great Northern prototype, and the Athearn 40-foot car follows a Union Pacific S-40-12,

but the roof panels aren't correct. Bowser makes models of Pennsylvania Railroad K-9 and K-11 cars. Central Valley makes an unpainted kit (less trucks and couplers) that nicely matches a common 40-foot Northern Pacific prototype.

Intermountain offers several variations of a 40-foot Santa Fe prototype (Sk-Q through Sk-U), and Red Caboose makes a Southern Pacific 36-foot S-40-5 car. Trix has a model of a UP S-40-12. Roundhouse makes a generic truss-rod car that would be appropriate into the early 1900s, and Mantua once made a car following a Reading prototype.

Sunshine Models, Westerfield, and Funaro & Camerlengo make several resin stock car kits for a variety of specific prototype cars. These take some time to make, but they build into good-looking, accurate models.

Choices in N scale are much more limited. The Micro-Trains 40-foot car is based on a 1940s New York Central prototype (rebuilt from older USRA boxcars). The Model Power 40-foot car follows a Union Pacific class S-40-12 car. Atlas, in the 1970s, offered a model of Northern Pacific's 86-foot Big Pig Palace car. Other available models don't follow specific prototypes.

AAR stock car classifications

The American Association of Railroads (AAR) classifies all freight cars to indicate their purpose and specific equipment or use. You'll find the classification stenciled on the car near the reporting marks and capacity data. Stock cars are classified S with several subclassifications. Here's a brief summary:

SC: Stock car having a convertible single or double deck, with roof, slatted sides, and side doors

SD: Stock car with drop-bottom doors in floor and a means of housing in the sides to close the slats, making a drop-bottom boxcar

SF: Double-deck stock car equipped with roof, slatted sides, and side doors

SH: Stock car specifically designed for transporting horses

SM: Single-deck stock car equipped with roof, slatted sides, and side doors

SP: Poultry car, with roof and wire-netting sides, fitted for shelves for shipping crates of poultry, with space allowed for feedbag and watering facilities

SPR: Combination poultry and refrigerator car, with one end equipped to haul live poultry and the other end refrigerated for carrying dressed poultry, eggs, butter, or other products

Meat refrigerator cars

From the late 1800s into the 1970s, fleets of insulated ice-bunker refrigerator cars (reefers) carried sides of beef as well as packaged, canned, and processed meat products from packing plants to distributors and buyers throughout the country. These cars, owned primarily by packing companies and other private owners, bore distinctive paint schemes and were mechanically different than reefers used for produce and other service. Although a few mechanically refrigerated cars came into use by the 1960s, the ice-bunker reefer served the meat industry until packing company traffic ended on the rails.

A string of new and old Swift refrigerator cars rests outside the company's Sioux City, Iowa, packing plant in 1951. *George Berkstresser, Lloyd Keyser collection*

This Chicago, Milwaukee & St. Paul refrigerator car, built in 1889, carried lettering for the Kansas City Packing Co. Note the archbar trucks and link-and-pin couplers on the 20-ton, truss-rod-underframe car. *David P. Morgan Library collection*

Early refrigerator cars

Through most of the 1800s, before mechanical refrigeration became practical, carrying finished meat products (or any type of perishable traffic) any distance was a challenge. By the 1880s, the railroads carried considerable livestock traffic—animals on the hoof. Through that period, it was much more practical to carry live animals to the areas where they would be locally butchered, processed, and consumed.

The lack of refrigeration meant during that period, most processed meat was cured and preserved—salted or smoked pork, for example. Raw, fresh meat, such as sides of beef or the resulting steaks, couldn't be moved even short distances without spoiling.

Attempts had been made to ship meat cooled with ice in standard boxcars, with little success. Several designers and companies worked to develop a practical car for transporting

perishables, but it wasn't until J.B. Sutherland developed and received a patent for the first true refrigerated railcar in 1867 that shipping meat (as well as fruits, vegetables, and other perishables) became practical.

Sutherland's car was the ice-bunker reefer. An ice tank inside each end of the insulated car provided the means for cooling the load. Vents atop each tank were designed to take in air while the car was in motion, pass the air over the ice, and circulate it throughout the car.

Various manufacturers worked to refine the car design, each putting a different twist on the basic car, resulting in several patent cars. The best known of these were the Tiffany car and the Wickes car. The Tiffany design featured an overhead bunker running the length of the car. The Wickes car had two metal ice bunkers, each with a woven galvanized metal exterior and additional metal fins attached to

increase the cooling surface. Cars following these designs often had the specific patent used noted in lettering on the car sides.

The various designs continued to evolve, and the refrigerator car proved successful. George Hammond is credited with making the first long-distance shipment of refrigerated meat in 1869. The growth of the packing companies around the Chicago Union Stock Yards in the late 1870s and 1880s led to larger-scale shipments, with Swift leading the way in 1877.

By 1880, just over 1,300 refrigerator cars were in use on American railroads. This number jumped to 23,000 by 1890 and 68,500 by 1900.

Private ownership

As refrigerator car design began to evolve in the late 1800s, railroads showed little interest in building or maintaining refrigerator cars for meat service. Railroads had already made a significant investment in stock

Refrigerator cars in service
1880: 1,310 (310 railroad-owned)
1890: 23,570 (8,570)
1900: 68,500 (14,500)
1924: 157,000 (33,000)
1932: 158,600 (46,000)
1941: 147,100 (22,100)

Armour operated the largest fleet of refrigerator cars into the early 1900s, leasing cars to others as well as using them for its own products. *Library of Congress*

cars. Refrigerator cars would increase shipping mileage of finished products, while decreasing the mileage of livestock shipping—which, since the animals hadn't been processed, resulted in more carloads and tonnage to haul.

Refrigerator cars were also more expensive to build and maintain than stock cars or boxcars, with their double-wall construction, insulation, ice tanks or bunkers, and need for constant ice replenishment. Add the fact they would spend half their time running empty, and railroads didn't want to make the additional investment for these specialty cars.

The result was that private owners, the packing companies and leasing companies, began building and acquiring large numbers of cars. Of the 68,500 reefers in service in 1900, about 54,000 were private-owner cars.

At that time, Armour was doing its best to corner the refrigerator-car market with a fleet of 20,000 cars. These cars not only served Armour's own needs but were leased to others. In fact, most of Armour's refrigerator cars were leased to haul produce in the West. By around 1920, multiple federal rulings required Armour to divest much of its fleet except cars for its own use.

As perishable traffic grew in the 1900s, some railroads began building their own fleets of refrigerator cars, including Northern Pacific and Illinois Central. A more common approach was for railroads to form separate companies to supply refrigerator cars, a tactic that placed fewer restric-

Mather cars, like this Morrell car with steel ends, enjoyed long service lives. The exposed channel side sill is a spotting feature of these cars. *William A. Raia collection*

tions on car usage. The largest of these were Santa Fe Refrigerator Despatch (later, Department) controlled by Santa Fe, and Pacific Fruit Express, a joint venture of the Southern Pacific and Union Pacific (and later Western Pacific). Other large private owners were Fruit Growers Express (Southern, Atlantic Coast Line, Baltimore & Ohio, and Pennsylvania) and its related companies, Western Fruit Express (Great Northern), Burlington Refrigerator Express (Chicago, Burlington & Quincy),

American Refrigerator Transit (Wabash and Missouri Pacific), and Merchants Despatch Transportation (New York Central).

The majority of these companies provided cars for produce service, but ART and MDT provided some meat cars. Other companies that provided meat reefers included General American (and subsidiary Union Refrigerator Transit), Mather Car Co., North American Car, and National Car Co. (a subsidiary of FGE/BREX).

Rath was a major user of Mather refrigerator cars. This steel-end car is at Chicago in 1966. *William A. Raia collection*

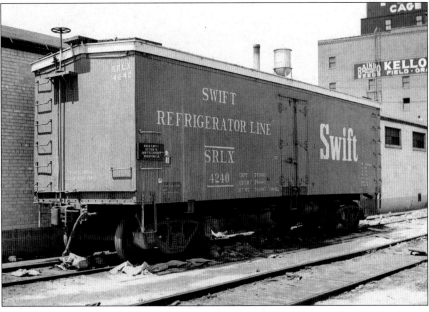

Number 4240 was typical of the Swift fleet into the 1950s. The 37-foot car has wood sides, ends, and roof and wears the company's red 1950 paint scheme. *David P. Morgan Library collection*

There was a big difference in how reefers in meat service were handled. Unlike cars hauling fruits and vegetables, meat cars remained in dedicated service to specific packing companies. For example, a car leased to Swift would never be found at an Armour packing plant or branch house.

On the other hand, a Pacific Fruit Express reefer might carry oranges from a California packing house to a Chicago wholesaler on one trip and then head to Texas to haul a load of lettuce to New York on its next run.

Ice-bunker car construction

Refrigerator cars of the late 1800s featured all-wood construction, following the same basic lines as contemporary boxcars. Underframes were typically wood with truss rods until the turn of the 20th century, when steel underframes began to appear.

Car sides had exterior vertical sheathing along with a separate interior wall. Insulating material filled the gap between the two walls. Hinged (swinging) double doors were standard on reefers by this time, as they could be thicker (insulated) and provide a tighter seal than the standard sliding door of a boxcar.

The major problem with early reefers was insulation. Materials initially used in the late 1800s (such as loose sawdust, powdered charcoal, and animal hair) simply didn't do a good job. They were poor insulators, and loose material tended to settle within the car sides, leaving large uninsulated areas in the walls. Water would sometimes seep through walls onto the insulation, which would also ruin its temperature-control qualities. Coupled with long transit times from the slow train speeds of the time, damage claims from spoiled products was a concern for shippers, builders, and railroads.

Improvements by the turn of the century included better insulation materials, such as blankets of felted hair, which wouldn't settle within walls.

Car length was typically 28 to 32 feet in the late 1800s, with 36-foot cars typical by 1900 and 40-foot cars by World War I. With the increased size came increased weight capacity as well: 20-ton capacity cars by 1900, 30-ton cars by World War I, and 40-ton cars by World War II.

As with boxcars, more and more steel came into use in the early 1900s. Steel underframes took over for wood underframes with truss rods (although some truss-rod cars were still built in the 1910s), and then steel framing for the bodies—even though sides and ends were often still wood. Steel roofs of various designs were common by the 1930s, again paralleling boxcar development. Steel-sided reefers began appearing in large numbers in the 1940s, but not necessarily for those in meat service.

Double swinging plug doors with a 4-foot-wide opening were standard into the 1950s, when sliding plug doors began appearing on new cars. Door width grew from 6 feet to 8 feet and wider to accommodate pallet jacks and forklifts.

Insulation kept improving. Celotex boards and fiberglass were common by the 1940s, with cork as well (especially under floors). The advent of polyurethane foam in the late 1950s was a tremendous improvement, and it enabled the growth of the insulated boxcar through the 1960s. These cars handled many shipments that once traveled by reefer. By this time, the ice-bunker meat reefer was on its way to extinction, phased out by mechanical cars and a shift to trucking.

Most of the major car manufacturers built meat reefers, including American Car & Foundry, General American, Mather, and Pacific Car & Foundry, along with major operators of cars (Swift and Armour) in the early 1900s.

Meat reefer differences

Refrigerator cars designed for meat service differed from cars in vegetable and fruit service in several ways, not all of them readily distinguishable from the outside. For starters, meat reefers were equipped with meat rails, ceiling-mounted metal railings that allowed sides and quarters of beef and hogs to be suspended on metal hooks from the ceiling. Door height was often lower on meat reefers to accommodate the meat rails.

A majority of reefers in meat service (about 60 percent) were equipped with brine tanks instead of standard ice bunkers at each end. These tanks held a mix of crushed ice and salt that provided a lower interior temperature than did ice alone. The tanks kept the brine from draining while the car

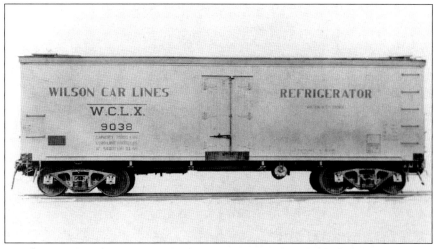

Wilson's car fleet comprised all 36-foot wood cars into the 1950s. This RSM car, with wood ends and roof, was built in 1945. It shows the company's simple post-billboard scheme. *Big Four Graphics*

This 38-foot Hygrade (Kingan) car, built in 1940, is a typical General American wood meat reefer of the late 1930s and early 1940s (earlier cars had straight side sills). It has wood ends and a steel-sheathed wood roof. *R.J. Burg collection*

was in motion—a major concern of railroads, as drippings from refrigerator cars could (and did) cause major corrosive damage to track and bridges. Brine tanks were more efficient than standard ice bunkers, although they were more expensive to install.

The brine tanks meant that meat reefers couldn't be used in ventilator service, unlike most cars in fruit and vegetable service. Cars used in ventilator service had hatches with adjustable latch bars, so the hatches could be propped open at various angles. With the hatches open, moving cars scooped in outdoor air that circulated through the car to

regulate the temperature. Doing this with loads of meat would dry out the hanging meat. The latch bars on cars with brine tanks didn't have the notches, so hatches were always closed unless the car was being cleaned or filled with ice.

In the 1940s, many produce reefers were equipped with air-circulating fans to regulate and even out the car temperature. Few meat reefers received these, as again the moving air would dry out the meat, and also because circulation wasn't as big a problem with a car of hanging meat as with a produce car packed tightly with crates.

Armour No. 687 is spotted at a Fayetteville, N.C., branch house in 1951. The 42-foot steel car, turned out by General American in 1948, was among the first steel meat reefers built. It wears the small Armour logo used until 1953. *Bob's Photo*

Meat reefers were typically shorter than produce reefers. Most cars designed for carrying meet were 36 or 37 feet long through the 1940s, while most produce reefers were 40 feet. Wood remained popular for sheathing on meat cars late into the 1940s, even when PFE, FGE, and other large fleets of produce cars had turned to steel sides. Many of these wood-sheathed cars remained in service through the 1960s.

The Association of American Railroads (AAR) specifies the mechanical designation for refrigerator cars, with several subclasses devoted to meat reefers (see the sidebar on page 31 for details). This designation is often stenciled next to the car's capacity information. Special equipment (meat rails and brine tanks) is often called out in additional stenciling on the car's sides.

Billboard cars

The 1910s and 1920s saw a growth in the number and style of so-called billboard refrigerator cars that used bright colors, large lettering, and logos to advertise shippers and specific products. Primarily operated by leasing companies, billboard cars could be found lettered not only for meat packers but also for dairies, creameries, breweries, and companies making and distributing beverages, produce, cleaning products, candy, dog food, and other products. (The photo on page 54 in Chapter 5 shows several examples.)

The era of the billboard reefer was relatively short-lived, coming to an end after a 1934 Interstate Commerce Commission ruling. It is often referred to as a ban on billboard-type cars, but that is really incorrect—it was more about illegal rebates.

During this period, railroads had fought the actions of many leasing companies, including the practice of lessors giving rebates to lessees based on cars' mileage income earned by the lessors. Part of the ICC ruling determined that the large billboard paint schemes amounted to illegal rebates to the lessees.

After the ruling (which took effect in 1937), freight cars used by a single shipper could have that shipper's name and emblem on them, but the scheme couldn't advertise any specific products.

Paint schemes were greatly simplified for the next 15 or so years, but by the late 1940s, bright

Wilson's later paint scheme, shown on 36-foot wood car No. 9242 in 1959, featured the company's W logo. Note the streaks down the car sides below the hatches and on the trucks. *J. David Ingles collection*

Meat reefer owners					
Company	1930	1940	1954	1962	1971
Armour ARLX	6,700	5,500	3,300	900	6
Armour TRAX (leased from ART)	—	—	598	594	—
Armour PCX (leased from PCL)	—	—	—	394	380
Cudahy Refrigerator Line	1,900	1,400	1,100	600	—
General American	700	8,100	5,300	3,300	960
Hormel [North American after 1930s]	120	—	—	—	—
Morrell (MORX, MRX)	660	640	800	—	—
National Car Co.	210	1,700	1,200	1,300	460
North American	1,600	2,000	2,000	6,800	4,700
Swift*	5,000	4,700	3,700	1,700	200
Union Refrigerator Transit	4,600	7,500	4,100	4,500	2,800
Wilson Car Line	2,400	2,400	1,500	1,100	900

*Swift cars are part of the General American fleet

Note: Many large and small packing companies leased cars from General American, Union Refrigerator Transit, and others, and had these cars painted in the packing company scheme. These cars are listed with the leasing company and not the packing company. All of the above cars may not have been in dedicated meat service at any given time, as the specific number of cars in lease service could vary widely from year to year (and even month to month).

National Car Co. leased meat reefers to a variety of small packers from the 1950s through the 1960s. Number 6820 is a 41-foot car with wood sides and ends and a steel roof. The car is not lettered for the lessee, but it is stenciled for return to a specific location. *J. David Ingles*

Swift's first large group of steel reefers were 39-footers built by General American starting in 1954. They were distinctive for their double latch bars, horizontal side seams, and tabbed side sills. *J. David Ingles collection*

colors and large logos were again appearing. These were primarily reefers in meat service, as they were among the only cars in dedicated service to specific shippers—unlike produce cars, which could be used by a number of shippers.

Mather cars

As with its stock cars, described in Chapter 2, Mather built and leased refrigerator cars for meat products. As with its stock cars, boxcars, and other equipment, Mather's formula was to use standard steel structural members and shapes in simple but sturdy construction.

Mather's initial reefers of the 1920s and '30s were 37-foot-long, 40-ton capacity cars with vertical wood sheathing on the sides and ends. The sides were distinctive (and provide a reliable spotting feature) because the steel chan-

nels used for side sills were visible below the wood sheathing on each side. The roofs were Mather's own interlocking-panel design.

Mather's meat reefers were equipped with standard ice bunkers at each end. These followed the company's economical ways, as bunkers were less expensive compared to brine tanks.

In the early 1940s, Mather bumped the length to 41 feet, retaining the wood sheathing. Immediately after World War II, Mather began building and rebuilding a modernized version of the company's earlier cars. These 42-foot-long, 40-ton cars still had wood-sheathed sides but used steel ends (improved Dreadnaught design with a three-over-three corrugation design) and steel roofs. These cars retained exposed-channel side sill of the earlier cars, making them easy to spot.

Mather's main customers were Morrell (starting in 1939) and Rath (starting in 1937). The cars were lettered in Morrell's and Rath's schemes with their own reporting marks. Of Mather's 1,300 cars in service in 1948, more than 700 were leased to Morrell and 500 to Rath.

Mather had 1,900 reefers in service by 1955, when the company (along with all of its cars) was acquired by North American. North American continued to lease Mather-built cars to various packing companies through the 1960s.

Armour car fleet

Packers developed their refrigerator car fleets in two basic ways: by buying cars outright (usually forming a car line as a separate arm of the company) or by leasing reefers on a long-term or short-term (per diem) basis. Each of the major packing companies handled this differently. Some owned as well as leased cars, and some packers leased cars from multiple sources. Cars on long-term lease

In the mid-1950s, National Car Co. rebuilt several 42-foot cars with steel sides and sliding plug doors. This one is lettered for Kahn's and stenciled for return to the company's Cincinnati plant. *R. J. Burg collection*

were often painted in a packer's paint scheme.

Armour historically had the largest reefer fleet among the major packing companies, even after divesting itself of the cars it leased to others. Armour owned the cars through its Armour Refrigerator Line (reporting marks ARLX).

The Armour fleet comprised various styles of wood cars, most 37 feet long, into the late 1940s. Most had been built in the 1910s, with 500 acquired in the early 1930s. Armour was one of the first packing companies to acquire a significant number of steel cars,

buying 2,000 42-foot cars in 1948 and 1949 built by General American (Nos. 1-1000) and American Car & Foundry (Nos. 1001-2000).

Along with its own cars, Armour also leased 100 rebuilt cars from Mather beginning in 1953 (with MARX reporting marks) as it began retiring its wood cars. Other leases followed, with 600 new steel cars from American Refrigerator Transit in 1954 (TRAX reporting marks) and 400 from Packers Car Line in 1957 (PCX), a subsidiary of National Car Co. The TRAX and PCX cars were built by Pacific Car & Foundry, and they could be

spotted by their overlapping horizontal seams that ran along the center of each side.

From the early 1900s into the 1930s, the company's wood cars carried bright schemes advertising its own meat products as well as dairy and other products it distributed. Cars carried Armour Refrigerator Line lettering on the left, with product names and logos on the right.

The scheme was simplified after the ICC billboard ruling, and a small logo (maroon with white ARMOUR lettering and a star) began appearing to the right of the door in the 1940s. Starting in 1953, this

AAR refrigerator car classifications

The AAR placed refrigerator cars in class R with several subclassifications. All are insulated house (enclosed) cars. Here's a summary:

RA: Brine tank reefer. Cars equipped with a brine tank designed for holding a mix of crushed ice and salt, usually without ventilating devices. Used chiefly for meats and packing-house products.

RAM: Brine tank reefer. Same as RA, but with ceiling-mounted beef rails for hanging sides of meat.

RB: Beverage, ice, water, vinegar car. Insulated car without ice bunkers and with or without ventilation devices.

RBL: Same as RB, but equipped with adjustable loading or stowing devices.

RCD: Solid carbon dioxide refrigerator. A car using carbon dioxide instead of ice as a refrigerant.

RP: Mechanical refrigerator. A car provided with apparatus (operated by power generated inside car) for furnishing protection against heat or cold.

RPA: Mechanical refrigerator. Same as RP, but with apparatus powered by direct mechanical drive from car axle.

RPB: Mechanical refrigerator. Same as RP, but with apparatus powered by electro-mechanical drive from car axle.

RPM: Mechanical refrigerator. Same as RP, but equipped with beef rails.

RS: Bunker refrigerator. Equipped with ice bunkers designed to hold chunk ice with or without ventilation devices.

RSM: Bunker refrigerator. Same as RS, but equipped with beef rails.

Dubuque Packing Co. leased cars from Union Refrigerator Transit, including this General American 39-foot steel car, shown in 1961. *J. David Ingles*

Swift's final paint scheme was silver with the oval Swift's Premium logo. Rust stains show from the ice hatches down the sides. *J. David Ingles*

logo was replaced by 36"-tall maroon ARMOUR lettering.

Swift car fleet

Swift, the other major packing company with a car fleet rivaling Armour's in size, rostered more than 6,000 reefers through the 1920s. The Great Depression led Swift to sell its entire fleet of cars to General American, which leased the cars back to Swift, giving the packer a leased fleet instead of direct ownership. The cars retained SRLX (Swift Refrigerator Line) reporting marks.

Most Swift cars through the 1940s were wood cars, 37 feet long, with wood ends and roofs. The ends of the cross bearers were visible below the car sides on most cars, but some series of cars had riveted plates at the ends of the cross bearers and bolsters. Door height ranged from 5'-9" to 6'-1". Almost 200 Swift truss-rod

Through the 1960s, Iowa Beef Packers leased cars from multiple sources, including Union Refrigerator Transit. This is a modern 43-foot steel car built by General American. *J. David Ingles collection*

cars (in the 10000 series) made it through the 1940s, and a few lasted into the 1950s.

The company had a few steel cars starting in the late 1940s but didn't have significant numbers until 1954 with the arrival of more than 770 39-foot cars built by General American. These cars, in the 15000 series, were distinctive because of their twin swinging doors with dual latch bars, tabbed side sills (the tabs covered the ends of the cross bearers), and horizontal side seams. Following that, many older wood cars were rebuilt with steel ends and sliding plug doors.

Into the 1930s, Swift's cars were among the boldest of the billboard cars, with large SWIFT lettering across the entire side, and lettering for specific products in areas inside the large letters. After this, the cars became much more sedate, wearing a simple yellow scheme with plain lettering.

The scheme became brighter in the late 1940s, as the company began adding its large red logo to the yellow cars. This was followed by bright red cars with large white Swift lettering in 1950, a scheme that continued through the decade. The company's final scheme was a switch to silver starting in 1959, with the Swift's Premium logo replacing the old rectangle.

In the 1960s, the company rebuilt a few of its steel cars, converting them into long 56-foot mechanical reefers in the 25000 series. By 1966, four cars had been converted.

Other reefer fleets

Wilson operated a reasonably large fleet of its own reefers, all 36-foot-long wood cars, into the late 1950s. The company finally began buying 42-foot steel cars in the late 1950s, acquiring about 800 of them and retiring the wood cars fairly quickly in the early 1960s. Wilson also bought 200 longer steel cars, 56-footers with sliding plug doors, in the early 1960s.

The cars wore a similar paint scheme throughout the post-billboard era, featuring orange sides with simple lettering. A small Wilson's logo—a *W* with *Wilson & Co.* spelled out in a box atop it—began appearing in the late 1940s. This was revised in the early 1950s with a larger logo, and by the late 1950s the *& Co.* was eliminated from the logo's lettering. In the 1960s, Wilson also leased some cars to smaller packing companies including Beefland International, Cornland, Needham, and Schuyler.

Cudahy had a significant fleet of cars in the early 1900s. It was well known for its billboard schemes in the early 1900s, including the Old Dutch Cleanser cars (shown on page 8). The scheme was simplified greatly in the 1930s and 1940s with orange sides and plain Cudahy Car Lines lettering, which remained into the 1960s.

Cudahy owned most of its cars but also leased cars as needed at various times from Mather (1930s–1942) and General American. All wore CRLX reporting marks. Most of its fleet was 37-foot wood cars with wood

Two of Morrell's Mather cars converted to mechanical refrigeration head east on the Illinois Central's Iowa Division in 1954. They were the first mechanical meat reefers. *Basil W. Koob*

North American built 75 53-foot steel meat reefers in 1961 for lease to Hormel. The cars had 6-foot sliding plug doors. *North American*

sides and ends, with orange or yellow sides and red ends.

Hormel began leasing cars from North American in the mid-1920s. Originally these cars had their own reporting marks (GAHX), but by 1940, the company's leased cars carried North American's NADX reporting marks. The fleet included older all-wood, 37-foot cars built in the 1920s, and by the 1950s, it included Mather cars as well as newer steel cars including some 53-footers.

Morrell had a fleet of its own wood reefers as well as cars leased from Mather, both the original 37-foot-long design with wood sides and ends and the 42-foot car of the late 1940s with steel ends and roof. In 1954, Morrell sold all its own cars to Mather as well. They ran with MORX (leased) and MRX (owned) reporting marks. Morrell also leased some cars from National Car Co., which had Morrell logos but NX reporting marks.

Morrell's earlier scheme featured orange sides with boxcar red ends and a large red Morrell logo. By the late 1950s, the company switched to a square blue logo with a red-and-white heart design and Morrell lettering.

Rath leased cars from Mather from the 1930s onward, operating early and late versions of the

---

Small lease fleets

Smaller packing companies leased cars from private refrigerator car lines, especially during the 1950s–1970s. Depending upon the lease length and fleet size, these cars carried reporting marks of the leasing company or had reporting marks for only their cars. Here's a summary of packing company logos that appeared on various lessors' cars during that period:

American Refrigerator Transit
Central Packing, Royal Packing

General American
Hygrade, Kingan

Merchants Despatch/Northern Refrigerator Car Line
Agar Packing, Jacob Dold Packing, Iowa Beef Packers, Minnesota-Iowa-Dakotas Packing, National Packing, Patrick Cudahy (Cudahy Brothers), Sioux City Dressed Beef

National Car Co.
E. Kahn's & Co. (EKSX), National Packing, Pepper Packing, Sioux City Dressed Beef

North American
Agar Packing, Hormel, Greenlee Packing, Kohr's

Union Refrigerator Transit
American Beef Packers, Black Hills Packing, Bookey Packing, Cudahy, Dubuque Packing, Dugdale Packing, Greenlee Packing, Hormel, Iowa Beef Packers, Marhoefer Packing, Needham Packing (Sioux City Dressed Beef), Oscar Mayer, Producers Packing, Raskin Packing, Rifkin & Sons

Mather wood reefers (710 in service in 1955, with RPRX reporting marks). Rath acquired several steel reefers from North American starting in the late 1950s. These cars were of various designs with swinging or sliding-plug doors and included some 55-foot cars.

The scheme was yellow/orange with a small version of the company's Indian head logo on the side in the 1940s. The logo became larger by the late 1940s, and by the early 1950s, a modernized version of the logo appeared in an outlined white rectangle.

Oscar Mayer had a variety of refrigerator cars through the years, leased from a number of car owners including Mather, MDT (MERX reporting marks), URTX, National Car Co. (OMX marks), and North Western Refrigerator Line (NWX). Cars had various schemes, most of which featured the company's logo with red lettering and a white background.

The end of the meat reefer era was filled with many colorful paint schemes, led by Iowa Beef Packers (later Iowa Beef Processors), which had cars painted in a variety of colors with the large IBP logo. Others included Agar, American, Dubuque, Greenlee, National, Needham, Pepper, and Royal.

Mechanical and converted cars

The first mechanical reefers began appearing in the late 1940s, but they were mainly developed for carrying frozen foods, which required much colder temperatures (below 0° Fahrenheit) than meat or produce.

A 1960 *Railway Age* article on mechanical refrigerators noted that of the 3,800 mechanical reefers then in service, only 46 primarily carried meat.

Morrell in 1952 had 25 Mather-owned cars rebuilt with Thermo-King mechanical refrigeration units (cars 2500-2524). However,

Wilson operated 200 56-foot steel cars in the 1960s. The cars had 6-foot sliding plug doors and wore the company's final paint scheme with the logo with large Wilson lettering. *J. Michael Gruber collection*

A key component of a meat refrigerator car is being equipped with meat rails. Sides of meat are hung by hooks along the rails. Note that the sides are stamped with identification and inspection labels. *New York Central*

Several of Armour's Packers Car Line leased reefers were fitted with end-mounted refrigeration units in the 1960s. This one, shown in 1970, shows the company's final scheme with large ARMOUR lettering. *J. David Ingles*

this was the Chicago, Burlington & Quincy, which bought 400 40-foot refrigerated meat trailers in the early to mid-1960s and added to the fleet in 1969 and 1970. In fact, in December 1969, the Burlington started a high-speed piggyback train dubbed the *Beef Express*, which departed Denver Friday afternoons bound for Chicago with trailers headed to Philadelphia and New York.

Over-the-road trucks won the traffic war, and most meat refrigerator cars were off the rails by 1971, as ice stations were removed from service and most produce cars were retired in favor of mechanical refrigerators.

Modeling

Red Caboose has made a Mather reefer in HO scale, Rapido has an HO model of a late-1930s 37-foot General American car, and Walthers makes HO models based on a General American steel reefer with overlapping side panels. In N scale, Micro-Trains has a wood refrigerator car that can stand in for a meat reefer.

To match specific prototypes, it's hard to beat resin kits. Sunshine and Westerfield have both offered resin kits for numerous meat reefers in HO scale.

these weren't intended just for cooling meat. Instead, they were designed to haul frozen concentrated juice back to the Midwest from Florida processing plants, getting more utilization from the cars.

Ten Burlington Refrigerator Express 55-foot mechanical reefers built in 1957 received meat rails in 1961. In the early 1960s, several hundred new, larger BRE mechanical cars received meat rails and were

given BRMX reporting marks (as opposed to BREX).

Armour retrofitted some 42-foot PCX cars with end-mounted reefers in the late 1960s, and other companies did similar experiments, including a Rath refrigerator car equipped with cryogenic refrigeration.

All was for naught, as meat traffic was heading to trucks. A last effort to claim meat traffic was with refrigerated piggyback trailers. One railroad that tried

The Chicago, Burlington & Quincy tried to recapture meat traffic in the 1960s with large (67-foot) mechanical refrigerator cars and 40-foot refrigerated trailers. *Chicago, Burlington & Quincy*

The heart of the livestock industry was the stockyard, where cattle, sheep, and hogs were brought to market, bought, and sold in large numbers. Rail operations were vital to stockyards of all types through the late steam and early diesel eras. Facilities ranged in size from a single-pen loading dock on a small-town railroad spur to huge union stockyards in large cities, where hundreds of thousands of animals sprawled across hundreds of acres.

All offer great modeling potential. In order to accurately model livestock operations, it's important to understand how each type of stockyard worked and how railroads handled traffic at each type of facility.

Side-by-side tracks enter the Omaha union stockyard. Loading platforms are at car-deck level, with portable ramps enabling loading and unloading of sheep and hogs from double-deck cars. *Library of Congress*

FOUR

Drovers load cattle into stock cars as a Chicago, Burlington & Quincy steam locomotive waits to move another cut of cars to the loading chutes. Crew members sit on the caboose cupola to watch the action in this 1941 view. *Hol Wagner collection*

Local stockyards

From the late 1800s through the 1960s, railroads maintained local stockyards along their routes in cattle-raising country. In the Midwest and West especially, it wasn't unusual to see pens in almost every town. The Chicago, Burlington & Quincy, for example, listed more than 660 stock pens along its lines in 1937; the Chicago & North Western had 200 such pens in 1940. In North Dakota in 1942, more than 90 percent of towns located on railroads had stock pens.

Even though the western United States is more noted for livestock traffic, plenty of animals began their journeys on eastern railroads as well, and pens could be found along railroads in most farming areas east of the Mississippi River. As an example, about 40 percent of Kentucky railroad towns had pens in 1942.

Most of these railroad-owned stockyards were relatively small. About 75 percent of them had four pens or fewer and could load 1–4 cars at a time with a siding or spur capacity of 6–12 cars. Larger stockyards at bigger towns could have 10–15 pens and be able to load 6 or more cars at a time with a siding capacity of 20 or more stock cars. The closer a town was to a major union stockyard, the less likely it was to have a pen.

These local stock pens were designed to allow farmers and ranchers a place to load their cattle for shipping to a union stockyard, packing company, or other buyer. Most were little more than a series of pens and perhaps a scale and a scale house.

Each railroad followed its own set of standard plans in building and maintaining local stockyards and pens, but most followed a similar design. The drawings on pages 40–41, based on the union stockyard at Sioux City, Iowa, can be used as a basis for modeling various stock pens and yards.

Pens were divided and surrounded by wood fences that stood 5–6 feet tall. Fence posts were placed every 5 to 6 feet, and fences were commonly horizontal 2 x 8s with a gap between each board. Wider boards, such as 2 x 10s, were sometimes used lower down on the fence. To protect animals, horizontal planks were used on

An empty stock car is spotted at a two-chute loading pen on the Chicago, Burlington & Quincy near Gillette, Wyo., in the 1940s. *Newberry Library*

both sides of fences that divided pens.

Loading chutes were located at trackside—one or two at a small yard, more at larger facilities. Depending upon the railroad or stockyard, chutes were 40–44 feet apart on center to match typical lengths of 36- to 40-foot coupled stock cars. For single-deck cars, wing gates swung out to meet the car sides. These gates often had sliding extensions that allowed them to better align with car doors.

The loading chute could be a single level, or it could have a permanent structure to allow loading the upper deck of a stock car as well, as was often done for hogs and sheep.

The Chicago, Burlington & Quincy's standard plans called for 8-foot-wide platforms, with the platform 3'-8" above rail height and the edge of the platform 6'-2"

from the track center line. The CB&Q plans show pens 96 feet square, which could be divided in half to form two separate pens. The Northern Pacific called for 24 x 40-foot pens, 8-foot-wide alleys, and 6-foot-wide platforms.

The size of a particular facility determined other support structures. A scale was important for weighing cattle before loading. By the 1940s, about half the western stockyards were equipped with a scale. The scale would be located in a small enclosure that protected the mechanism or in a scale house, which could also contain a small office.

Larger installations might have one or more covered pens. These pens were used mainly for hogs that must have shade if kept outdoors for any period of time because they are sensitive to heat and sunlight. Other structures on site would include a shed or barn

for hay and straw, a water tank and well house (and possibly a windmill), and storage sheds for sand, which was used for bedding.

Many towns and small cities featured livestock sale and auction barns that were locally owned by a municipality or private owner or company. In addition to a building for showing animals and conducting auctions, these sites commonly had a series of storage pens. Although much of the business at these facilities catered to local buyers, with most livestock arriving and leaving via truck, many were located on rail spurs to facilitate long-distance shipping.

Rest stations

Railroads also operated larger stockyards that were designated as rest or feed stations. Their purpose was to provide a place to off-load livestock as mandated by

Overhead view of a standard receiving pen

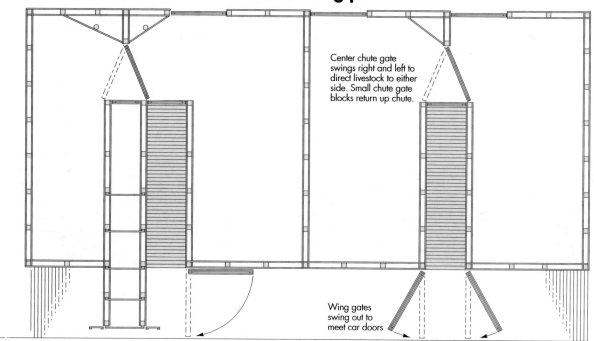

Center chute gate swings right and left to direct livestock to either side. Small chute gate blocks return up chute.

Wing gates swing out to meet car doors

Trackside view

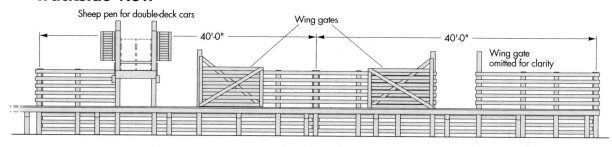

Sheep pen for double-deck cars

Wing gates

40'-0"

40'-0"

Wing gate omitted for clarity

Hay is stacked by pens at the union stockyard in Denver in 1939. *Library of Congress*

Side view of loading chutes and platforms

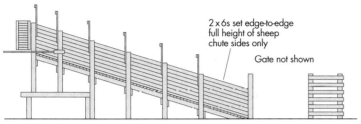

2 x 6s set edge-to-edge full height of sheep chute sides only

Gate not shown

High-level chute with counter-balanced footplate

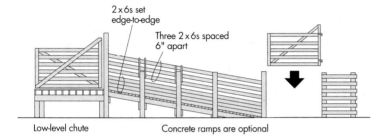

2 x 6s set edge-to-edge

Three 2 x 6s spaced 6" apart

Low-level chute

Concrete ramps are optional

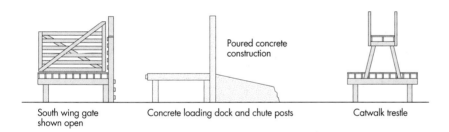

Poured concrete construction

South wing gate shown open

Concrete loading dock and chute posts

Catwalk trestle

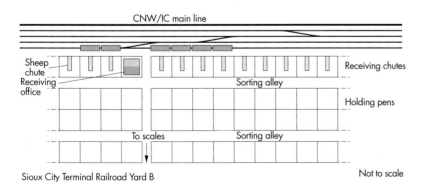

CNW/IC main line

Sheep chute

Receiving office

Receiving chutes

Sorting alley

Holding pens

To scales

Sorting alley

Sioux City Terminal Railroad Yard B

Not to scale

The drawings shown here, based on the union stockyards in Sioux City, Iowa, can be used as a modeling guide. Most prototype pens followed similar construction styles.

Modeling pen fences

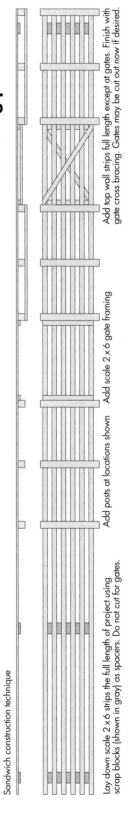

Add top wall strips full length except at gates. Finish with gate cross bracing. Gates may be cut out now if desired.

Add scale 2 x 6 gate framing

Add posts at locations shown

Sandwich construction technique

Lay down scale 2 x 6 strips the full length of project using scrap blocks (shown in gray) as spacers. Do not cut for gates.

A big group is on hand to watch as a carload of steers is loaded onto a Santa Fe stock car at a local pen. *Santa Fe*

A semi brings a load of cattle to a railside loading pen at San Angelo, Texas, in 1939. *Library of Congress*

Market terms for livestock

Cattle

Bulls: noncastrated males, generally sold for breeding

Cows: females

Steers: castrated males, usually destined for the beef market

Butchers: cattle suitable for processing into cuts of meat

Cutters: lower-grade cattle

Canners: lowest grade, not suitable for processing into cuts of meat; often used for processed (or canned) meat products

Sausage/bologna bulls: cattle not carrying enough meat for processing into cuts, but suitable for sausage products

Stockers: young cattle requiring more fattening by grazing or feed

Feeders: cattle ready to be placed on feed diets

Hogs

Barrows: castrated males

Sows: females

Gilts: young unbred sows

Butchers: mature barrows or sows ready for market

Skips: low-weight or low-quality hogs

Sheep

Springers: young lambs

Yearlings: young sheep (1–2 years old)

Wethers: castrated males

Culls: low-quality lambs or sheep

Canners: low-weight sheep not suitable for processing into meat cuts

the 28-hour law (see Chapter 6 for details). In brief, the law required that cattle not be confined to stock cars for longer than 28 consecutive hours. Sheep were an exception and could be confined for 36 hours. Other livestock could also be confined for 36 hours if the owners signed a waiver.

In 1936, this Northern Pacific stock train is loading at the cattle pens at Medora, N.D. When the drovers are done loading the car, the train will move additional cars into position. *Library of Congress*

Because many long hauls took longer than 28 or 36 hours, railroads established these rest and feeding stations. In addition to its 600-plus local stockyards, the CB&Q operated 21 larger rest stations in the late 1930s, with 9 designated as main feed yards.

These rest and feed stations were large operations. The CB&Q's Galesburg, Ill., facility had 163 pens with a capacity for 236 cars. Many of these rest stations were designed just to feed and water animals, but others included larger pens or grazing land for longer-term storage of cattle. Sellers could contract with the railroad to hold animals for a period of time in hopes that market prices would rise.

Open pens were combined with covered pens and barns to hold various types of animals. Water and feed (hay or grain as specified by the shipper) were provided to livestock, with the cost billed to the owner. Large barns and sheds held hay, straw, and feed.

Incoming cars were unloaded, the stock cars were cleaned, and then livestock was reloaded to continue their journey. (These operations are explained in Chapter 6.) Along with the inbound and outbound stock cars, the stations would receive shipments of hay and feed in boxcars, or depending upon the location, by local trucks.

Union stockyards

The largest livestock operations were at union stockyards (so named because ownership was usually formed among a group of smaller stockyards and other owners). Located in many cities across the country, union stockyards provided a central place for sellers to bring their livestock, with a variety of buying agents on site, including packing companies, breeders, feedlot operators, and others.

For most of the livestock-by-rail era, Chicago's Union Stock Yards was the largest operation in the country. Opened in 1865, the stockyard was initially formed by the nine railroads that then served the city, with the goal of consolidating all stock operations at a single location. The stockyard served the city's packing companies, but more importantly at the time, it provided a place to buy and sell cattle that were arriving from the West and headed to eastern markets.

Chicago's operation was a huge success, and by 1880, the stockyard covered more than 350 acres and handled 2 million head of livestock a year. Its pens could hold

Cowboys drive cattle to pens at Brawley, Calif., in 1942. *Library of Congress*

Hay and feed could be brought to stockyards by truck or rail, depending upon location. This scene is on the Northern Pacific at Dickinson, N.D., in 1936. *Library of Congress*

10,000 cattle, 120,000 hogs, and 5,000 sheep, and it was served by more than 1,500 stock cars a day.

Over the next 25 years, other cities soon formed their own large union stockyards to serve the growing western market. Major stockyards included Kansas City (opened in 1871), St. Louis (1872), Indianapolis (1877), Omaha (1884), Sioux City (1884), Denver (1886), South St. Paul (1886), Fort Worth (1887), and Sioux Falls (1889). The table on page 53 shows the total number of livestock passing through major stockyards during the 20th century, and the map displays the locations of the largest stockyards.

As explained in Chapter 5, meat packers began building plants next to union stockyards in many cities as the stockyards provided a ready supply of livestock—an

Cattle wait in pens at the union stockyard in Winnipeg, Manitoba, in 1949. Most eastbound Canadian Pacific livestock trains stopped here for feed and rest. *Canadian Pacific*

arrangement that worked well for buyers as well as sellers through the 1940s.

The drawing on page 50 shows the Omaha union stockyard as it was in 1925. A union stockyard covered hundreds of acres. Most of the area was taken up by cattle pens, which included open as well as covered pens, and buildings. Each pen had a unique address, and the whole complex was connected by alleys that functioned like streets in a city.

Rail spurs entered the stockyard at various points. Loading and unloading platforms were at car-floor level with either permanent second-deck ramps or portable ramps that could be wheeled into position for upper-deck loading or unloading.

Large structures and sheds or barns were maintained for storing straw for bedding as well as hay and grain for feed. Other structures housed supplies, machine shops, and equipment such as tractors, loaders, sweepers, and other implements necessary for cleaning pens or hauling hay, straw, and feed to pens.

A livestock exchange building was a focal point of a union stockyard. These multistory structures were often distinctive works of architecture and stood out dramatically among the sea of surrounding cattle pens. The exchange building held offices for commission agents, livestock buyers, regulatory officials, and stockyard officials.

Union stockyards employed hundreds, and sometimes thousands, of workers. Workers unloaded and loaded stock cars and trucks, led cattle among pens and scales, cleaned pens, and handled feed and water.

Along with railroads, trucks also served union stockyards. Whether cattle arrived by rail or truck largely depended upon the distance traveled. As highways improved and trucks became larger, trucks carried more and more livestock. At Chicago, for example, by the early 1950s, trucks carried nearly 90 percent of livestock to the stockyard.

Buying and selling livestock

When a rancher or farmer wanted to sell an animal or group of animals, he had several options. He could sell it directly to a buyer—a packing company, feedlot operator, or another farmer or

Buyers on horseback roam the alleys among pens of Omaha's union stockyards in the early 1940s. Many cattle would eventually make their way to the looming Cudahy plant or other neighboring packing plants. *Library of Congress*

Union stockyards had extensive docks for loading and unloading trucks. This is Buffalo, N.Y., in the late 1940s. *New York Central*

A load of steers from the West walk off a Santa Fe stock car at the Buffalo, N.Y., union stockyard in 1945. Stock cars would sometimes roam far from home rails, just like other freight cars. *New York Central*

rancher. This was commonly done for local butchers (in which case transport would usually be by truck) or packing companies located in smaller cities that didn't have next-door access to a union stockyard. These packers relied heavily on local animals carried by truck, but they could receive livestock by rail as well.

A more common arrangement was for the livestock owner to contract with a commission company. Commission companies had agents at union stockyards, and the agents would contract with owners to sell livestock for them.

Prototype example for shipping cattle

Here's a typical scenario for shipping cattle. Rancher A in western Nebraska decides to sell two dozen full-grown steers. He contacts his commission agent and then notifies the local Union Pacific station agent that he has cattle to ship. The agent orders a stock car and arranges a date for pickup. A local freight delivers the car to the UP's local cattle pen before the loading date.

The morning of the shipment, Rancher A loads the cattle into his truck for the ride to the UP cattle pen. Upon arrival, he backs the truck to the unloading gate. The cattle are weighed and moved to the appropriate pen. Other cattle from multiple sellers may be arriving at the same time. On this day, two other sellers also have animals being loaded, so three stock cars are in position at the loading chutes. Rancher A and another rancher each have con- tracts with commission agents in Omaha, and a third rancher is selling a group of smaller feeder cattle (cattle not yet fully grown which are ready to be matured on grain feed) to a feedlot operator in Iowa.

Rancher A gets his steers into a stock car an hour in advance of the local freight train, and the UP agent finishes waybills for each car (more on those in Chapter 6). As the doors close, the clock begins ticking on the 28-hour law. Livestock can only be confined in stock cars for 28 hours (36 if a waiver is signed—see Chapter 6 for more details). The time limit shouldn't be an issue for this trip, but the rancher has signed a waiver authorizing the cattle to be confined to 36 hours if necessary. This will save him feed costs if there is a delay.

Because these are small shipments going relatively short

The livestock exchange building was a focal point at most union stockyards. It housed offices for commission agents, buyer agents, and other officials. This is the exchange building at Kansas City in 1906. *Library of Congress*

The Santa Fe maintained a small two-pen, one-chute stockyard at Milan, Kan. *Santa Fe*

Sheep are unloaded from a double-deck Milwaukee Road stock car at South Omaha in 1938. *Library of Congress*

distances, no representatives of the sellers (drovers) will be traveling with the animals. Handling will be done under contract at the destination stockyards.

The local freight train arrives, picks up the three stock cars at the head end, and is soon on its way to the UP's large classification yard at North Platte, Neb. Upon arrival, all three cars are quickly pulled from the local freight and, along with eight other stock cars, added to the head end of a priority eastbound freight train. The train is soon on its way, covering the almost 300 miles to Omaha in about 9 hours.

Upon arrival at Omaha, all 11 stock cars are immediately pulled out of the train. The car heading to the feedlot in Iowa is hauled to

the interchange track with the Chicago & North Western, which will carry it to its final destination. Two other cars of steers are routed to the Chicago, Burlington & Quincy, as the animals' owners have a contract with a commission agent at the St. Joseph, Mo., stockyard. Those cars will soon be on their way and should make the 140-mile trip without hitting the 28-hour limit.

The remaining seven stock cars are routed to the Omaha union stockyard, where they are all unloaded. Stockyard employees (called *key men*) take the animals from each car and lock them in temporary holding pens (catch pens). Representatives of Rancher A's commission company check the waybills, settle the railroad freight charges, and take the

animals to pens assigned to the commission company. The steers are then watered and fed.

Livestock were generally held for at least a day to allow them to regain weight lost in transit. They might remain in their pens for two days or longer until a sale is made. Buyers—representatives of local packing companies, feedlots, and others—dealt with commission sales agents and bid on various lots of cattle. Offers were by hundred-weight (per hundred pounds).

In Rancher A's case, a buyer from Cudahy made an offer for the steers, which the commission agent didn't accept. A short time later, an Armour buyer inspected the cattle and made an offer, which the agent accepted. Once the bid was accepted, a commission company yardman took the

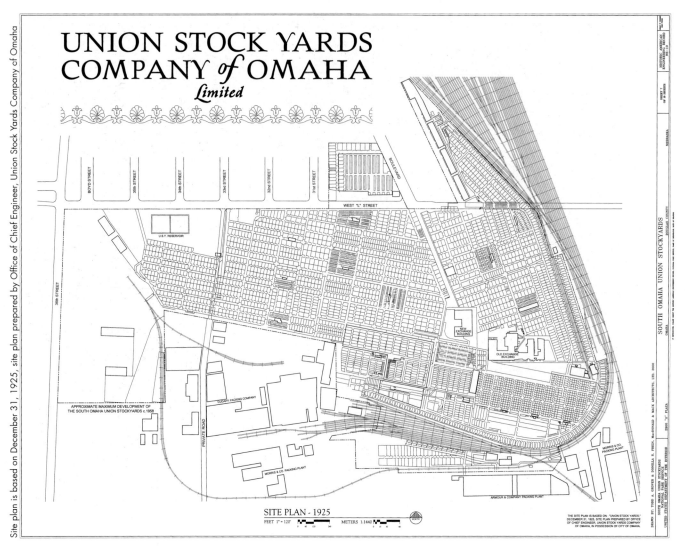

UNION STOCK YARDS
COMPANY *of* OMAHA
Limited

BOYD STREET

36th STREET

34th STREET

33rd STREET

32nd STREET

31st STREET

BOULEVARD

WEST "L" STREET

36th STREET

U.S.Y. RESERVOIR

NEW EXCHANGE BUILDING

OLD EXCHANGE BUILDING

APPROXIMATE MAXIMUM DEVELOPMENT OF THE SOUTH OMAHA UNION STOCKYARDS c.1958

CUDAHY PACKING COMPANY

PRIVATE ROAD

MORRIS & CO. PACKING PLANT

MORRIS & CO. PACKING PLANT

ARMOUR & COMPANY PACKING PLANT

SITE PLAN - 1925
FEET 1" = 120'

METERS 1:1440

Historic American Engineering Record map of the Omaha union stockyard. *Historic American Engineering Record, National Park Service, drawn by Todd A. Grover and Donella R. Frein, MacDonald & Mack Architects, 2000*

Hogs were kept in covered enclosures, often multistory buildings such as this one at the union stockyard in South Omaha. *Historic American Engineering Record*

Union stockyards had several scales and scale houses, where livestock would be weighed to complete a purchase. This one is at Omaha. Gates were located on either side of the scale. *Library of Congress*

animals to a scale. As soon as the weight was registered, the sale was complete.

Stockyard employees then moved the now-sold cattle to catch pens. The new owner would then pick up the cattle. Rancher A's former cattle would be picked up later that day by an Armour employee, who would drive them directly to the killing floor at the neighboring Armour packing plant.

Other buyers might route cattle to their own pens for eventual shipping by stock car or truck to a distant feedlot, ranch, stockyard, or packing plant.

All union stockyard operations were cash only, settled on the day of transactions. Stockyard, commission, and buyer agents operated by handshake and

spoken word. Although paperwork was generated to verify weight and sales details, the industry was largely known for its trust in handshake deals.

Declining numbers

The number of local railroad-maintained cattle pens and stockyards slowly declined from the 1930s onward, as trucks began hauling animals longer distances. Small two- to four-pen railroad facilities were abandoned in favor of larger stockyards.

Receipts at large union stockyards began falling in the 1940s and dropped dramatically from the 1950s into the 1960s as packing companies began abandoning their large traditional plants that neighbored union stockyards. New plants were built

in rural areas near feedlots, and packers began buying animals directly instead of through agents.

The number of feedlots—where large numbers of cattle were raised in a small space and fed grain (instead of traditional grazing) grew dramatically in the 1950s and 1960s. Advances in antibiotics and their widespread use greatly reduced the risk of disease and epidemics among large groups of animals in close quarters, making feedlots economical. Also, Americans tended to favor the flavor of marbled grain-fed beef over that of grass-fed steers. By the early 1960s, about two-thirds of transported cattle went to feedlots before slaughter.

Another factor was the growth of livestock auction markets that

Small sheds located throughout the stockyard served as offices for commission agents. This is the Star Commission Co. shed at South Omaha. *Library of Congress*

Cleaning the stockyard was a constant process. A fleet of tractors and sweepers continually scooped used bedding and manure from pens and alleys. *New York Central*

brought buyers and sellers together without commission agents. Local markets as well as large stockyards increased auction sales over time.

Without a ready neighboring market for animals, many large union stockyards closed their gates. All of Chicago's major packing plants that neighbored the stockyards were closed by 1960, and Chicago Union Stock Yards closed in 1971. Omaha became the largest union stockyard in 1955, foretelling a western shift in the market. However, three of Omaha's four major packing plants closed in the late 1960s: Cudahy in 1967, Armour in 1968, and Swift in 1969. Omaha lost its largest status in 1973 to South St. Paul but remained open until 1999—long after rail operations had ceased.

Major U.S. stockyards, 1930s–1940s

1. Chicago
2. Cincinnati
3. Denver
4. Fort Worth
5. Indianapolis
6. Kansas City
7. Louisville
8. Milwaukee
9. Ogden
10. Oklahoma City
11. Omaha
12. Peoria
13. Pittsburgh
14. St. Joseph
15. St. Louis
16. San Antonio
17. Sioux City
18. Sioux Falls
19. South St. Paul
20. Wichita

Top 10 union stockyards
Total livestock received (in millions)

	1904		1934		1954		1964	
1	Chicago	15.3	Chicago	13.0	Chicago	5.0	Omaha	5.2
2	Kansas City	5.4	Omaha	7.1	Omaha	4.9	South St. Paul	4.1
3	Omaha	4.9	Kansas City	5.9	South St. Paul	4.7	Chicago**	4.0
4	St. Louis	3.7	South St. Paul	5.8	St. Louis	3.7	Sioux City	3.8
5	St. Joseph	3.0	St. Louis	5.5	Sioux City	3.5	St. Louis	3.1
6	Indianapolis	2.0	Sioux City	4.7	Indianapolis	2.9	St. Joseph	2.4
7	Cincinnati	1.5	Denver*	4.6	Kansas City	2.3	Kansas City	2.1
8	Sioux City	1.5	St. Joseph	3.5	St. Joseph	2.2	Indianapolis	2.0
9	Fort Worth	1.2	Ogden*	2.6	Fort Worth	1.8	Sioux Falls	1.9
10	Denver	0.9	Pittsburgh	2.3	Sioux Falls	1.3	West Fargo	1.1

*Denver and Ogden figures unavailable for 1954 or 1964
**Estimated

Source: *America's Historic Stockyards: Livestock Hotels* by J'Nell L. Pate

By 1970, most rail operations at stockyards were finished, other than special shipments for large customers. The remaining union stockyards scaled back operations and were served by trucks only.

Modeling
Walthers offers a stockyard kit in HO scale (No. 933-3047), as well as scale cattle. Campbell Scale Models has an HO stock pen, loading pen, and unpainted cattle. Other HO items include loading ramps from GC Laser and Classic Miniatures, hogs from Boley, cattle from Dyna Model Products, and sheep from Merten and Preiser.

In N scale, The N Scale Architect offers a stockyard kit, and GC Laser makes a stock car loading ramp.

Cattle pens, both large and small, can be built from scale stripwood. Start with scale 4 x 4 or 6 x 6 posts with 2 x 8 fencing material. You'll need a considerable amount of wood for even small facilities, but construction is not difficult. Look at photos and scale drawings for details. For the best look, be sure to stain all of the wood appropriately before beginning.

FIVE

A long string of Swift billboard refrigerator cars, many of which proudly advertise individual products, stands in front of Swift's Sioux City plant in 1917. The classic multistory design is typical of packing plants built in the late 1800s and early 1900s. *Sioux City Historical Museum*

Packing plants are the focal point of the meat industry. Dressed meat as well as processed, packaged, and canned meat products are produced and shipped to outlets across the country. Through the 1960s, railroads played a key role in getting packing house products to market.

The packing plants themselves, as well as the branch houses that served as outlets for products, are ideal candidates for modeling. From the late 1800s onward, packing plants grew to become huge operations, with individual plants processing thousands of animals a day. We'll take a look at how the industry evolved, the process itself, and some ideas on modeling a packing plant or branch house.

A handler drives several steers up a ramp toward the kill floor at the Armour plant in Sioux City in the 1940s. The ramp comes from the neighboring union stockyard. At right, it goes under the parking area. *Sioux City Historical Museum*

A Milwaukee Road switcher moves a cut of refrigerator cars at the Hormel plant in Austin, Minn., in 1942. Most of the reefers are wood-sheathed cars leased from North American. *David P. Morgan Library collection*

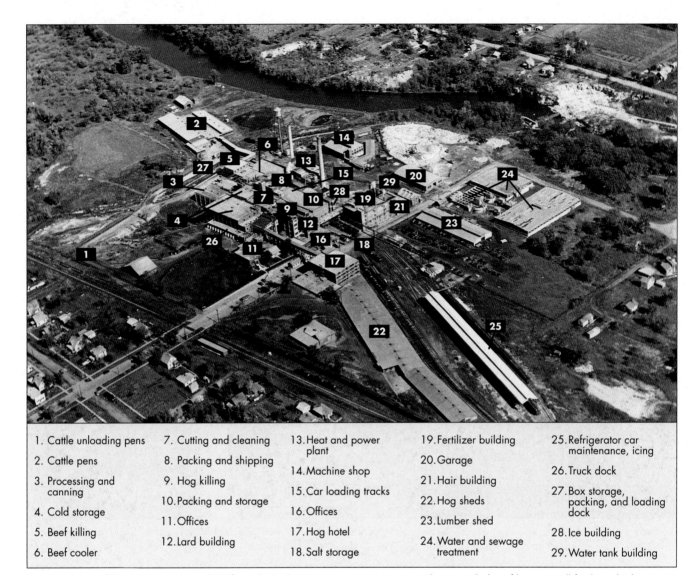

1. Cattle unloading pens	7. Cutting and cleaning	13. Heat and power plant	19. Fertilizer building	25. Refrigerator car maintenance, icing
2. Cattle pens	8. Packing and shipping	14. Machine shop	20. Garage	26. Truck dock
3. Processing and canning	9. Hog killing	15. Car loading tracks	21. Hair building	27. Box storage, packing, and loading dock
4. Cold storage	10. Packing and storage	16. Offices	22. Hog sheds	
5. Beef killing	11. Offices	17. Hog hotel	23. Lumber shed	28. Ice building
6. Beef cooler	12. Lard building	18. Salt storage	24. Water and sewage treatment	29. Water tank building

This aerial view of the Armour packing plant (formerly Decker) in Mason City, Iowa, provides a good idea of how a small facility is laid out. Note that the hog storage areas (17 and 22) are covered to keep the hogs out of the sun. *David P. Morgan Library collection*

History

Through most of the 1800s, the business of butchering and processing meat was a small-scale operation. Slaughterhouses operated in most areas of the country, using animals raised locally. A lack of electricity and refrigeration meant butchered animals had to be either consumed right away or preserved, which was usually done by smoking, salting, or pickling. Much of the industry was seasonal, with many plants working strictly in the winter.

The invention of mechanical refrigeration helped revolutionize the meatpacking industry. In 1859, Ferdinand Carré of France developed a vapor-compression system using ammonia that made large-scale building refrigeration practical.

Breweries were the first industry to extensively apply refrigeration, beginning in the 1870s, and commercial ice-making operations soon followed suit. Meat packers were a bit slower to catch on, but by the 1880s, the use of refrigeration was widespread in the industry.

Refrigeration allowed packing plants to become large-scale operations. Instead of being preserved, beef and pork quarters and halves could be held in large refrigerated warehouses and processed or shipped as needed.

Ice-bunker refrigerator cars, which grew in numbers from the 1880s and 1890s onward, allowed these large packing companies to ship dressed meat and meat products all across the country.

Chicago became known as the world's leading meatpacking center as several packing companies set up shop next to the city's union stockyard, led by Swift in 1875 and followed shortly by Armour, Nelson-Morris, and others.

By 1900, the major packing companies had built large plants near union stockyards in many other cities, mainly in the Midwest and Plains states where animals were raised. Being located next to

Several billboard refrigerator cars wait outside the Cudahy Packing Co. Sioux City plant in the early 1900s. The main part of the plant stands about six stories tall. *Sioux City Historical Museum*

A mix of older wood reefers and new steel cars mingle on the cleanout and ready tracks at Swift's Sioux City plant in 1954. A lone car in the old yellow scheme stands out among a sea of cars in the red 1950 paint scheme. *George Berkstresser, Lloyd Keyser collection*

a union stockyard provided a ready supply of livestock.

By the 1910s, the industry was dominated by what became known as the Big Five packing companies: Armour, Cudahy, Nelson-Morris (purchased by Armour in 1923, which made the group the Big Four), Swift, and Schwarzchild & Sulzberger (which became Wilson & Co. in 1917). All five companies operated plants in Chicago and also had large packing plants in other cities.

These companies dominated the market, controlling more than two-thirds of domestic meat production in the early 1900s. In fact, their hold on the market was too great, and in the 1910s the companies were charged with conspiring to fix prices and control the market. As a result, the packers were required to divest themselves from stockyard and retail ownership, as well as other ancillary businesses not directly related to packing plants.

Along with the Big Four, many other significant packing companies developed through the 1900s including Cudahy Brothers (later Patrick Cudahy, a separate entity from Cudahy Packing Co.), Decker, Dubuque, Hormel, Oscar Meyer, Kahn's, Morrell, and Rath.

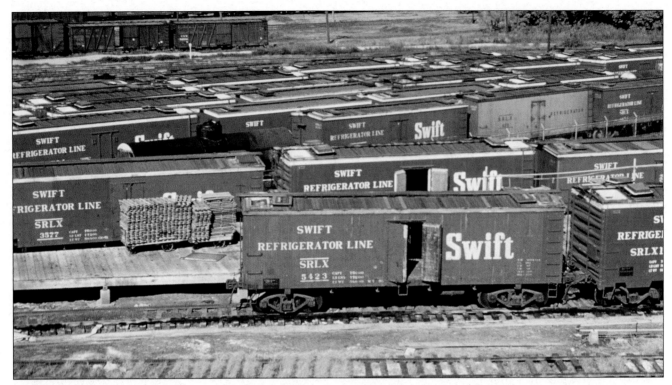

Stacks of floor racks await on a platform next to Swift cars with their doors and hatches open. A car cleanout or repair area presents some interesting modeling possibilities. *George Berkstresser, Lloyd Keyser collection*

Packing plants

Packing plants were pioneers in the assembly line process—although it's often noted that the term *disassembly line* is more accurate for slaughterhouses. In fact, Henry Ford, the man most-often credited with developing and perfecting the moving assembly line, wrote in his autobiography that it was seeing Chicago packing plants in action that inspired the development of the processes at his auto plant.

The process of slaughtering and processing animals is not a pleasant one, and into the 1900s, the conditions at most large plants were, in a word, repulsive. Upton Sinclair's *The Jungle*, when first published in 1906, led to public outcry and eventual passage of the Meat Inspection Act and the Pure Food and Drug Act of 1906. Although conditions improved, the very nature of turning animals into meat is not for the faint of heart.

Packing plants divided the work into several individual steps, so each worker was responsible

for a single operation, such as performing a specific cut. This enabled workers to be trained in a smaller number of tasks, which sped up the overall process.

Traditional packing plants were generally multistory structures that featured top-down operations. The smaller the plant size, the fewer buildings it contained and the more operations that were combined. Large plants had separate buildings for most procedures.

The photo of the Armour plant in Mason City on page 56 provides a good overview of how a smaller packing plant was laid out. There are separate kill floors for each type of animal (beef and pork at this plant), with separate pens for each.

Plants had separate buildings or areas for meat processing, storage, packaging, and shipping. A non-refrigerated warehouse or storage areas held packaging materials, salt, feed, and other items that did not require cooling. A power plant is marked by a tall smokestack.

Other manufactured products (lard in the case of the Armour Mason City plant) were made in their own area or building. The larger the plant, the more likely that by-products were manufactured into new products on site. Smaller plants shipped by-products to rendering plants.

On-site water treatment plants became standard at medium- to large-size plants in the early 1900s. Shops for general maintenance and refrigerator-car repairs were located on plant grounds. An office building, an ice storage building, and a reefer-icing dock could also be found.

Plant operations

The process starts with acquiring the animals themselves. Particular plants may specialize in just beef or pork, or others may slaughter several types of animals, which could also include sheep.

Large packing plants located next to a union stockyard acquired most (often, all) of their animals through the stockyard. Packing company buyers at the yard could purchase animals as needed based on daily needs. These animals were then driven on the hoof directly from the stockyard to the plant.

Smaller packing plants had holding pens at railside, and they also had truck unloading docks and pens that may or may not be connected to the railroad pens.

Cattle and sheep were usually kept in open pens, but hogs needed protection from sunlight. They were kept in covered pens or, at some packing houses, in multistory structures with open sides.

Cattle pens at packing plants were not meant for long-term storage. Feeding was done as needed, but it was counterproductive to feed animals that will be slaughtered in a short time. Plants tried to avoid keeping more than a day's production of animals on hand at any given time.

Animals were led into the plant as needed. At older multistory plants, the killing floor was at the top level, with ramps leading from the cattle pens or stockyard.

The process proceeds very quickly once it starts. The animal is killed and blood drained from the carcass. The hide is then removed and the carcass cut into parts.

All parts of every animal are used in the process, the most important being cuts of meat. Through the 1950s, most cattle were prepared as dressed beef (carcasses cut into halves or quarters) that was then shipped to branch houses. Depending upon the packing plant, meat could also be processed into sausage or other specialty cuts or preserved (smoked, cured, or canned) as products such as ham and bacon.

Specific types of animals were purchased depending upon need. For example, plants would buy low-grade cutters and canners at prices much lower than higher-grade butchers to make processed

The Swift packing plant in Sioux City exemplifies classic packing-house architecture, with multistory brick construction. The roof sign is a great detail to model. *Courtesy Sioux City Historical Society*

Packing companies established branch houses in large and small cities across the country. This early 1900s postcard shows a new Armour branch house in Johnstown, Pa. *Jeff Wilson collection*

This postcard shows another Armour branch house, in Gary, Ind. An unloading platform along a rail spur is at right. The rail above the platform is for hanging dressed meat. *Jeff Wilson collection*

Swift's branch houses followed a distinctive architectural style. This is the Swift branch house in Caldwell, Idaho, around 1940. *Library of Congress*

meat products. (See page 42 for animal classifications.)

The scale of operations at these facilities was huge. Even a small packing plant would process 1,500 or more animals in a day. Large plants processed more. For example, according to a 1930s visitor's guide to the plant, Swift's Chicago plant could process 750 hogs per hour, 800 sheep per hour, and 180 steers per hour.

Branch houses

The branch house system was developed in the late 1800s by Gustavus Swift. The basic system worked well and was soon copied by other meat packers as an ideal method for distributing their products. The basic idea of a branch house was to avoid whole-salers by providing an outlet to sell directly to butcher shops, meat markets, restaurants, and other end consumers.

Packing companies established branch houses in large and small cities across the country, with

more than 500 total branch houses in operation by Big Five companies by 1900. Swift had 189 branch houses and Armour had 152. This number kept increasing. By 1948, Armour had more than 250 branch houses, and Swift had around 400 at the system's peak.

The branch house system was driven backward through the process by its final consumers. Salesmen at branch houses would take orders from customers. Based on their requirements, each branch house would communi-cate a daily order via telegraph to the company's central office or to a specific plant. As orders arrived and were assigned to specific packing plants, each plant would then know how many animals needed to be slaughtered that day and in what form (halves, quar-ters, special cuts, packaged items, or other products).

Going back another step, livestock buyers at stockyards for each packing plant then knew exactly how many of each type of

animal (as well as what grade and specific breed) would be needed each day. They could adjust their bids and make purchases appropriately.

Products to fill orders were then loaded on refrigerator cars, each car bound for a specific branch house. Large branch houses might receive one or more cars per day, and smaller branches perhaps one or two per week. Branch houses had delivery trucks adorned with the meat packer's name and paint scheme. Custom-ers could also pick up their orders at the branch house.

Branch houses performed various functions depending upon the specific company and size of the operation. Most performed additional meat cutting and grind-ing to customer's orders. Some performed other processing such as sausage making. Most sold addi-tional products affiliated with their owner's brand. (The following section on by-products gives an idea of the scope of these products.)

Physically, a branch house looked like a cold-storage warehouse. Those of the large packers were typically built to common designs, as shown by the Swift example on the opposite page or the Armour branch houses on page 59.

Many were multistory designs. A rail spur and unloading dock would be located on one side of the building and a truck loading dock on another side. The building would also contain offices, a refrigerated storage area, and whatever other space was needed for processing.

Associated with branch houses, peddler cars served markets and towns that were too small for a branch house. Salesmen took orders from retailers in a series of towns along a rail line, and a packing plant loaded a refrigerator car with all the orders. The car would then be routed to each town and parked on a team track, where customers would pick up their orders.

Peddler cars were common from the late 1800s (Armour started the practice in 1887) through the 1920s. They served 16,000 towns in the 1910s. The practice then faded as highways improved and trucks began delivering products to markets at longer distances from branch houses.

By-products

Every part of each animal is used. The meat itself was the primary product, but not all meat might be fit for human consumption. Lower grades of meat and cutoffs were used for processed meat products or pet food.

The most valuable by-products were animal hides, which would be sent to tanneries to make leather and leather products. Great care was taken to not damage the hides when removing them from carcasses.

Another by-product was wool, which was used for clothing,

In these photos, workers unload dressed beef from a wood-sheathed Cudahy refrigerator car in New York City in the late 1940s. The meat hooks travel along an overhead rail on the dock, making it easier to move the meat into the branch house. *Both photos: New York Central*

blankets, insulation, and other products. Hair from other animals was often used for making insulation, felt, bristles on brushes, and stuffing for pillows and mattresses.

Fat and tallow were reclaimed for use in many products. Edible fats were used to make lard, margarine, candy, and gum. Inedible fats went into cosmetics

and soap, grease and other lubricants, glycerin, candles, and various chemicals.

Tankage refers to blood and bone meal and rendered meat, and it is often destined to become animal feed. Other products that have historically used animal by-products include pet food (organs), gelatin (bones, hooves),

Refrigerator cars could carry a combination of dressed meat and cases of products. Here, a worker loads a reefer at Armour's East St. Louis plant in the 1940s. *Newberry Library*

glue (bones, cartilage, hooves), fertilizer (dried blood, bones, hooves), buttons and handles (bones, horns), racket strings and sausage casings (intestines), pharmaceuticals (glands, blood), and neatsfoot oil, a leather conditioner (bones).

Outbound reefers

The most obvious rail operations at a packing plant were the outbound loads of meat and meat products. Because fresh meat is highly perishable, it received priority handling both at the plant and by the railroads, as Chapter 7 explains.

Refrigerator cars must be cleaned and inspected for defects as they arrived at the plant. Several tracks were dedicated to this function. The cars were then precooled with a load of ice prior to loading. Large plants had their

own icing stations for doing this, while a small plant might rely on a nearby ice service or railroad-owned icing station.

Once precooled, the reefers were moved to position on loading tracks at the plant. Loading progressed quickly. Dressed beef, hogs, and sheep were hung by hooks in refrigerated storage rooms at the plant. The dressed halves and quarters moved along railings to the loading dock, where workers carried them into the reefer and hung them from the meat rails mounted to the ceiling.

Cases of finished product were loaded by hand or hand truck and then stacked inside the car. As the above photo shows, cars could be loaded with a combination of dressed meat and cases of product, if that's what the destination branch house had on order that day.

Cars would be spotted for loading overnight or in early morning, with outbound loads sealed and ready to pull by mid- or late afternoon. If more cars were to be shipped than there was room for on the loading tracks (for example, if 40 cars were to be shipped but the dock had room for just 15), loaded cars would be pulled to a ready track and empties immediately spotted in their place at the dock.

Local switch jobs or freights assigned to the plant would bring the loads to a nearby yard, where they would be blocked in trains. Depending upon the size of the plant and number of neighboring packing plants, railroads might switch cars two or three times a day. Many union stockyards and neighboring packing plants were served by terminal railroads. (The Sioux City Terminal Railway and South Omaha Terminal Railway are two examples.)

Railroads handling a lot of meat traffic combined cars from multiple packers in trains that would be due out of town in the evening (see Chapter 7). The number of cars a plant loaded in a day would depend upon the size of the facility, the amount of processed products and dressed meat to be shipped, and the amount of products shipped by truck or other methods.

A 36- to 42-foot ice-bunker reefer could hold roughly 400 hanging beef quarters or the products from 100 steers. A 1,000-pound steer will yield about 550 pounds of dressed meat, and a typical hog will yield around 150 pounds of dressed meat.

Other rail operations

Animal hides were another important outbound load. These were shipped to tanneries for conversion to leather. A typical steer hide weighs about 60 pounds, and they are graded and sorted by size, type, and quality. When hides are removed from the

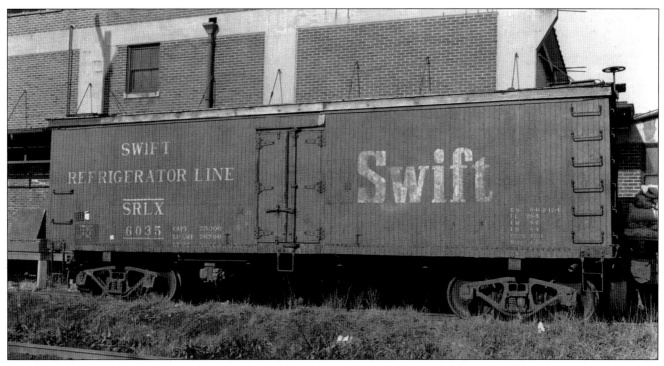

One of Swift's 36-foot wood refrigerator cars in the bright red scheme is spotted for unloading at the company's branch house in Mineola, N.Y., in 1954. *Bob's Photo*

carcass, they are salted (to preserve them), sorted, and tied in bundles.

Hides could be shipped by truck or boxcar, depending on the distance to the tannery. Boxcars used in hide service were stenciled with HIDE LOADING ONLY or similar lettering. They were generally old railroad-supplied cars at the end of their service lives because, once they carried a load of smelly hides, they weren't good for much else. One car of hides was shipped for every 20–25 outbound reefers.

Fats, tallow, and tankage were shipped to rendering plants in tank cars. Swift and Armour both operated fleets of tank cars to service their packing plants. As with its reefers, Swift owned its own tank cars until 1931, when the company sold all its cars to General American, which leased the cars back to Swift. The tank fleet bore SWTX reporting marks, and had 499 8,000- and 10,000-gallon cars in service in 1948. Other tank cars were leased from several sources, either by the packing company

or rendering company. Two or three tank cars were shipped for every 20 reefers of meat.

Incidentally, tank cars of by-products is one part of the meat industry that still has significant rail traffic. Many modern plants are rail-served and have loading stations for tank cars on their way to rendering plants. These cars are often stenciled for tallow and fat service.

Other outbound loads depended upon the amount of non-meat production done at a packing plant. Bones, dried blood, hooves, horns, and other by-products would go to a rendering plant, which was often nearby—meaning trucks did the hauling, not the railroad. Shipping this material a long distance by rail simply wasn't cost effective.

Inbound loads

Packing plants also received significant inbound traffic. Stock cars of animals seem to be the logical number-one load, but this often wasn't the case. As noted earlier, packers located next to union stockyards typically

obtained their livestock on the hoof directly from the stockyard, herding it over directly to the plant's kill floor.

Smaller plants, or those located away from union stockyards, received livestock by a combination of stock cars and trucks. Trucks brought in animals from local farms, ranches, or feedlots. (Chapter 2 provides some guidelines for the number of various kinds and sizes of animals that would fit into a stock car.)

Again, depending upon the specific products produced, a plant would receive boxcars of packing material (boxes, crates, tin cans), carloads of product additives and other ingredients (bags of salt, seasonings), boxcars and tank cars of chemicals (barrels in boxcars), and boxcars of wood (for smoking meat at some plants). If a packing plant had an on-site power plant, inbound cars of coal or fuel oil would also be received.

Plants could be served by in-house switchers, but as noted earlier, many union stockyards

Hide-service cars were generally old boxcars at the end of their service lives. This single-sheathed Chicago Great Western car is stenciled for hide service only. *Bob's Photo*

Standard tank cars carried inedible and edible fats, tallow, blood, and tankage from packing plants to rendering plants. *David P. Morgan Library collection*

were served by terminal railroads. These would serve the neighboring packing plants as well.

The new era of packing

As mentioned in Chapter 1, the market began changing dramatically in the 1940s and '50s. The old packing plants built around the turn of the 20th century were becoming obsolete, and companies began building new plants in small towns and rural areas closer to the sources of livestock. This meant fewer cattle went

through union stockyards and there was less long-haul cattle traffic for railroads.

Chicago, once the nation's leader in meatpacking, saw all of the Big Four packers leave the city during the 1950s. In 1954, Cudahy was the first to close its Chicago plant, followed by Swift in 1958 and Armour in 1959.

The new generation of packing plants were sprawling single-story structures with all operations on one level that featured much more mechanization.

The late 1950s saw the birth and growth of the interstate highway system, along with the growth of trucks. The 40-foot semitrailer became legal nationwide in 1957. Mechanical refrigeration units for trailers had become common, reliable, and cost effective by that time, and trucks began hauling more and more finished meat products.

The biggest change came with a new breed of packing companies in the 1960s. Iowa Beef Packers (IBP) built its first plant in Denison, Iowa, in 1961. The company's goal was to locate plants near feedlots and buy cattle directly. The company divided jobs into more steps than did traditional packing plants, which allowed more untrained (and thus cheaper) labor to fill most jobs.

The major change IBP implemented was processing much of its meat into final cuts at the packing plant instead of selling dressed meat to a butcher, who would do the final cutting. Meat would be cut, packaged, and put in cases (refrigerated or frozen), ready for retail sale. This tactic worked well and coincided with the emergence

Final delivery was usually made by truck (a panel truck here), as shown in this 1930s scene in Omaha, Neb. *Library of Congress*

of large supermarket chains and chain fast-food restaurants.

Although **IBP** shipped some of its products, including standard dressed meat, by rail during the 1960s, the shift toward using trucks was well underway by then. Other regional packers began copying IBP's business model, and most traditional packing plants near large union stockyards closed by the end of the 1970s.

During the 1960s, railroads and shippers made some attempts to switch to mechanical refrigerator cars, but trucks were already taking too much traffic for this move to be profitable. The end of the ice-bunker refrigerator car era around 1971 effectively meant the end of transporting fresh meat by rail.

Modeling

You have several options in modeling packing facilities. To convincingly model a large packing plant requires a great deal of space. You can selectively compress a facility by locating it along a backdrop, but you'll still need several tracks to serve inbound and outbound traffic.

Modeling a smaller stand-alone plant is reasonable, but this can still take a lot of room. One example of such a plant, with its operations labeled, is the Armour Mason City plant on page 56.

A simpler approach is to model a branch house. Since these were found in large and small cities across the country, they're appropriate for almost any region or railroad. You might even have room for two or more, perhaps from multiple packing companies.

Walthers offers an HO model of a small classic packing plant (No. 933-3048). It could be used as a starting point for building a larger plant when combined with other large brick structures. Walthers also offers a cold storage warehouse (No. 933-3020) that could be used to model a branch house. Heljan offers similar structures in N scale.

Using modular structure components in HO or N would also be a good way to build either a packing plant or branch house. Brick components from DPM or Walthers are appropriate for an older plant. Concrete wall sections from Great West Models would be suited for a more modern plant.

Alpine Division Scale Models (formerly Suydam and California Model Co.) offers a wood-and-cardstock HO craftsman kit for a Swift packing plant (actually a branch house) that makes a good base for constructing a more detailed model.

SIX

Chicago, Burlington & Quincy No. 5606 leads a priority freight, including a block of stock cars at the head end, through Bishop, Ill., in 1954. *Robert Milner*

Although never a major traffic source as compared to grain, coal, or other industries, livestock traffic was still a significant revenue source for railroads. It peaked at 1.5 million carloads per year during the 1920s, declined through the Depression, and came back a bit afterward. But by the 1950s, stock traffic was dwindling. A shift to trucks and the growth of feedlots, together with a changing market that saw packing companies move closer to feedlots, doomed long-distance livestock traffic by rail.

Many facets of livestock operations can be modeled. You can model local stock pens or a large stockyard. You can operate a string of livestock cars at the head end of a priority freight train or model the operations of a dedicated livestock special, giving it priority handling as it crosses your layout.

A long string of stock cars is tucked in behind Santa Fe 2-8-2 No. 4085 on train 44, the *Perishable Express*, near Henrietta, Mo., in 1946. Behind the stock cars are 33 refrigerator cars. *Robert R. Malinoski*

Early stock traffic

Much livestock traffic through the mid-1800s involved relatively short-distance transports. As the country expanded and people began moving west—and railroads expanded with them—ranchers began raising livestock on the wide-open ranges of the Plains and Southwest. Since the largest consumer markets were still in the East, getting these western animals to consumers became a long-distance challenge.

As Chapter 1 notes, the first large cattle drives brought animals to railheads in Kansas and other western states. Railroads then carried the animals to Midwestern and eastern markets. Since meatpacking was still largely a seasonal, regional business, due to limited use of refrigeration,

animals had to be shipped close to where they would be consumed. This meant long trips: it was a slow 1,500 miles from Abilene to New York City via Chicago.

As transport distances increased in the late 1800s, slow train speeds and long transportation times caused big losses during transport. Animals were confined to cars for several days at a time—five or six days was common for a Chicago to New York City trip in the 1880s. Cattle lost weight from a lack of food and water, and many weaker animals died en route. Other animals were gored, fell, were trampled, or suffered broken bones and other injuries from car movement.

So-called "humane" or palace stock cars, with built-in water and feed troughs, were an early

attempt to solve these problems. They were not extremely successful—feed troughs were not consistently refilled, water often didn't last long, and the constant motion of cars was not conducive to keeping animals healthy. The only way to get animals to market healthy was to limit the amount of time they spent in a stock car.

28-hour laws

In 1869, Illinois became the first state to require railroads to unload animals at periodic intervals: 5 hours of rest after 28 hours in a car. A similar federal law followed in 1872, but these laws had too many loopholes and exceptions, and enforcement was lax. Time laws and restrictions were virtually unenforced through the turn of the century.

A South Omaha Terminal Railway switcher adds cars onto the meat train transfer at the Terminal yard on a darkening October evening in 1957. The cars will soon become part of Illinois Central's hot CC-6 perishable train to Chicago. *William D. Middleton*

Workers unload a St. Louis-San Francisco stock car at a Santa Fe feed and watering station in Shawnee, Okla., in 1946. *Preston George*

This changed in 1906 with the national adoption of an amended version of the 28-hour law. In a nutshell, the law stated that animals could not be kept aboard a stock car longer than 28 hours. At that point, animals must be unloaded into "properly equipped pens" for rest, water, and feeding for at least 5 hours.

Exceptions included sheep, which could be confined up to 36 hours if the 28th hour would require unloading at night, or other animals if the owners signed a release that allowed them to remain aboard for 36 hours. Another provision allowed a longer period "if animals cannot be unloaded due to accidental or unavoidable causes that could not have been anticipated with due care."

The net effect of the law on rail operations was to reinforce livestock as a priority commodity.

Railroads began earnestly trying to handle stock to avoid the expense and time of having to stop trains to unload and rest cattle.

Traffic patterns

The overall pattern of stock traffic in the United States was firmly established by the turn of the 20th century. The basic flow for live-stock was moving from grazing lands in the West and Southwest to stockyards and packing companies in the Midwest or East. As population increased in California and elsewhere on the West Coast, a smaller flow of livestock moved from the Plains states to packers on the West Coast.

As previously explained, the largest stockyard and concentration of packing companies was in Chicago, with other centers in the Midwest stretching from Sioux Falls and South St. Paul southward through Omaha, Kansas City, and Fort Worth. The map on page 53 in Chapter 4 shows major U.S. stockyards in the 1930s and '40s.

Although the majority of stock shipped by rail eventually reached one of these centers, the reality was that stock could be

A Chicago, Burlington & Quincy switcher starts the livestock hauling process by setting out a pair of empty stock cars for loading at a Wyoming stock pen in 1955. *William A. Akin*

shipped from anywhere to anywhere. Stock could be shipped from ranch to ranch, from ranch to open grazing land and back, or from ranch to feedlot.

Along with animals destined for slaughter, other traffic included livestock that needed more maturing, dairy cows, and bulls and boars for breeding.

In the fall, cattle and sheep were often shipped from blizzard-prone areas to warmer grazing lands in the south; in the spring, this was repeated as livestock returned north. This traffic was substantial through the 1930s, until trucking took over most short- to mid-range hauls.

North of the border, livestock in Canada followed a similar pattern. It was typical for cattle to be moved from grazing land in Alberta and Saskatchewan to market centers and feedlots in the east. This traffic was heavily seasonal with a lot of fall traffic.

Western railroads carried the

Annual livestock carloadings			
Year	Loadings	Year	Loadings
1925	1,550,000	1950	491,000
1930	1,280,000	1955	447,000
1935	712,000	1960	259,000
1940	685,000	1965	125,000
1945	894,000	1968*	74,000
*1968 was the last year the AAR tracked livestock carloadings as a separate category			

bulk of livestock through the 20th century. The chart above lists the total number of stock carloadings for several years, and the chart on page 71 shows the top 15 carriers (by carloads hauled) for 1951. The Santa Fe and Union Pacific were always at the top of the list. The top railroads east of the Mississippi predominantly hauled stock from Chicago to New York City and other East Coast markets.

Home road cars tended to dominate most stock train consists, but stock cars often went off line, just like other types of cars. If you scan the photos

throughout this book, you'll see many examples of this.

Handling stock cars

Moving livestock by rail presented several challenges to railroads. As the ultimate perishable product, live animals were picked up as close to the actual loading time as possible. The 28-hour law dictated operations, so trains carrying livestock were on priority schedules and often rated new motive power.

Loaded stock cars were handled as fast as possible, but a resulting problem was that fast train speeds

The Sioux City Terminal Railway served the city's union stockyard as well as the surrounding packing companies. Here, an 0-6-0 goes to work with the Cudahy and Armour plants in the background. *Henry J. McCord*

meant more rocking and jarring of cars, causing potential problems from slack action running in and out. This is why loaded stock cars were almost always carried at the head end of a train. This afforded the least slack action and the smoothest ride.

Railroads emphasized smooth train handling with loaded stock cars, spelling this out in special rules or employee guidebooks. Stock cars weren't sent through hump yards, and kicking or dropping (switching maneuvers involving cars rolling free from a locomotive) loaded stock cars was not allowed. For smooth switching and coupling, loaded stock cars were to have air brakes cut in whenever moved.

Having stock at the head end also made it easier for terminal switching. Stock cars were cut out and forwarded or switched immediately upon arrival at a yard or terminal. Caboose crews appreciated head-end handling as well, since it minimized the aroma.

Livestock was moved in priority freight trains whenever possible. Stock was often combined with other perishable traffic (meat and produce). Also, until the decline of stock traffic in the 1950s, dedicated all-livestock trains were found on many railroads. These trains could be run as multiple sections of an existing freight or as an extra for which dispatchers would give priority handling. These trains often weren't long, 25–50 cars, to avoid slack action on the rear of the train.

Empty stock cars would be routed back to their home railroads in general manifest freight trains, and they could be found anywhere in a train.

Local operations

Train operations started with delivering stock cars to stockyards or pens. Cars were cleaned after each trip: old straw, sand, and manure would be shoveled out, and cars would be power-washed

(or steam cleaned) to remove residue. Lime was often applied to each deck.

Fresh bedding would be placed prior to loading. This was typically straw or a 2"–3" layer of sand, but it could also be sawdust or—into the early 1900s—cinders. The cleaning and prep work could be done at the site where the car would be loaded or, if the loading site was not equipped to do so, at a larger facility.

Livestock operations often started at local railroad-owned stockyards (see Chapter 4), where local trains would pick up loaded cars. These local stockyards varied widely in size, from having a few pens to those with a dozen or more. Depending upon the number of cars to be loaded and the spacing of the pens, multiple switching moves would be needed to pick up cars.

This local or stock day train would carry the cars to the nearest classification yard or connection with the main line.

Among the best-known dedicated livestock trains was the Union Pacific's Daylight Livestock, train 299. It's heading west behind a matched set of new Alco FAs in 1951. *Donald Sims*

Stock cars would be combined with others and placed in a time freight or, if traffic warranted, a dedicated all-stock train.

If traffic necessitated, a dedicated train might be coordinated to pick up stock from several stockyards along a route on a single day. Operations would progress with the train picking up cars at each town along a line (described in Chapter 4). This was especially common on branch lines in the West through the 1930s.

Empties would have been left in place earlier by a local freight, and cars would be prepped and bedded. The stock train would be run as an extra, so on the day of the special, a locomotive would receive a train order to run (with a caboose) to the end of the branch, or most distant stockyard on the route, and begin picking up cars.

The train might have to conduct multiple switching moves at each facility. For example, if a town had a one-chute stockyard

Top livestock carloadings by railroad, 1951	
1. Union Pacific	92,500
2. Santa Fe	86,400
3. Burlington	68,500
4. Southern Pacific	53,600
5. Chicago & North Western	51,000
6. New York Central	45,104
7. Milwaukee Road	43,800
8. Missouri Pacific	35,800
9. Pennsylvania	33,300
10. Rock Island	30,300
11. Nickel Plate Road	25,700
12. Baltimore & Ohio	22,900
13. Rio Grande	21,800
14. Northern Pacific	21,600
15. Great Northern	20,200

but three cars were to be loaded, the train would have to reposition each car for loading.

If a town's stockyard had three loading chutes, but the chutes were on 40-foot centers (designed for 36-foot cars) and 40-foot cars

were supplied, the train would likewise have to reposition the cars for loading.

A stock extra could be called to handle as few as a dozen cars or as many as 30 or 40. The key was the number of towns and stockyards

The narrow-gauge Denver & Rio Grande Western did a lot of livestock business. Here, several stock cars are loaded with sheep in Silverton, Colo., in 1942. The cars are switched one at a time to the single loading chute. *William Moedinger*

that had loaded cars to ship, and whether that was any more than the line's regular local train could handle expeditiously. Because a local freight would have additional switching work to do at several towns en route, the stock cars could be delayed by several hours—a critical factor in complying with the 28-hour law.

Local operations changed a great deal from the 1930s into the 1940s as roads improved and trucks became larger. Livestock owners began trucking livestock longer distances—often directly to a larger union stockyard.

To combat this, in the 1930s, many railroads began providing a livestock pickup service. They contracted with local trucking companies to bring livestock to a larger railroad-owned pen for further transport by rail. However, in 1940, an ICC decision barred railroads from doing this. By this time, many small-town two- and

three-pen stockyards were being abandoned.

By 1940, about 60 percent of total livestock traffic went by truck. In areas where stock transportation distances were relatively short (New England and Northeast), this percentage was much higher (other than through traffic heading to New York, Boston, or other large cities). This was reflected by the relatively low number of stock cars owned by eastern railroads. It was in the West, where transport distances were long, that railroads kept a significant share of stock traffic.

Basic stock operations

Most railroads that handled a lot of livestock published instructions on their care for employees. The process started with the shipper notifying the local agent and the agent ordering the number and type of cars needed. Cars were inspected to make sure there

weren't any exposed nails or bolts, broken boards, or defects that could injure cattle.

The waybill (prepared for each car) listed the car's reporting marks and number, the number of animals loaded into each car, and the type and condition of the stock. (A sample stock waybill is shown on the opposite page.) In addition, a separate livestock contract was signed by the owner.

The waybill shows the time the car was loaded, the car's routing, whether the shipper signed a 36-hour waiver (a separate form), and the intended stops for rest and feed. Blanks in the waybill were filled out any time the stock was unloaded and reloaded, including times and charges for feeding.

The railroad usually provided and prepped the bedding material. Sand was typically used in hot weather, especially for hogs since the sand could be wet down

R. A. O. A. Standard Form No. 101

A. B. & C. Ry. Co.

LIVE STOCK WAYBILL

ORIGINAL

Stop this Car at	For	Gross Weight of Oar and Contents for Engine Rating	Date	Waybill No.
St. Louis, Mo.	Feed & Water	30 Tons	August 13th.	Local 29

Car Initials and Number	Transferred to		From — (Station)	(State)
A.B.C. 4329	At		Number (381) Pronto, Ala.	

To— (Station) (State)

Chicago, Ill.

Full Name of Shipper, (Origin and Date, Original Car, Transfer Freight Bill and Previous Waybill Reference When Rebilled)

P. Shannon,
Pronto, Ala.

Route (Show Each Junction and Carrier in Route Order to Destination of Waybill. Indicate by Check Mark Whether Shipper's or Agent's Routing.)

Shipper's Routing

Agent's Routing

Car Ordered. To be Filled In when Different Car is Furnished

Consignee, Address (Final Destination and Additional Routing)

Great Lakes Com. Co.
Chicago, Ill.

Length of Car		Kind		Height	
Ordered	Furnished	Ordered	Furnished	Ordered	Furnished
36 ft.	36 ft.				

Time Loaded __9 A.M.__ Date __August 13th.__

Was an Attendant In Charge? (Yes or No) __Yes__

Was Car Bedded by Carrier? (Yes or No) __No.__

Has 36-hour request been signed and filed at point of origin?

(Yes or No) __Yes__

Government Certificate No. __B 27301__ Attached.

Indicate by Symbol in Column Provided ★ how Weights were Obtained.

R—Railroad Scale

A—Weighing Bureau or Agreement.

T—Tariff Classification or Minimum.

S—Shipper's Agreement or Tested Weight.

E—Estimated (Weigh and Correct)

Weighed

At _____

Gross _____

Tare _____

Allowance _____

Net _____

Number of head and description of stock. Note—Ordinary live stock unless otherwise specified.	Weight	★	Rate	Freight	Advances	Prepaid
C/L Cattle (20 more or less)	20,000	T		107.50		

Destination Agent's Freight Bill No.

Feeding and Rest Record

PLACE	Pen No.	Unloading Record			Reloading Record			Dead & Crippled Removed	Feed		Price Cwt. or Bu.	FOR ADDITIONAL CHARGES SEE		
		Date	Time	Count	Date	Time	Count		Amt.	Kind		Amount	Waybill No.	Date
Total (For Destination Agent's Use Only)														

All Yard Stamps To Be Placed on Back of the Waybill in Spaces Provided

Junction Forwarding Agents Will Show Junction. Stamps in the Space and Order Provided Below:

Destination Agents will Stamp Herein the Date Reported

1st Junction	2nd Junction	3rd Junction	4th Junction

Additional Junction Agents' Stamps Must be Shown on Back of Waybill in Spaces Provided

This generic livestock waybill, suggested by the Railway Accounting Officers Association, was followed by most carriers. The waybill lists information including the car's reporting marks and number, routing, weight and condition of animals, and feed and rest stops.

Railroads moved a lot of animals to and from grazing lands in fall and spring. These sheep are grazing along the Union Pacific main line near Lookout, Wyo., in 1952. *Donald Sims*

Several steers look on as a car inspector checks the journals on a stock car at the South Omaha Terminal yard in October 1957. The car is one of several that will soon be heading eastbound on a hot livestock/meat train on the Illinois Central. *William D. Middleton*

For large shipments of stock by a single owner, representatives of the owner (drovers) could accompany the shipment. The drovers would then handle livestock whenever they were unloaded and reloaded. For smaller shipments (less than a carload), the cattle, once in transit, would usually be handled by stockyard employees or railroad employees at railroad-owned rest and feed stations. The shipper would then be billed for this service.

Rest and watering stations

Stops at feed and watering stations were built into train schedules for livestock traffic. When a train arrived at a rest station, drovers or stockyard workers unloaded the cars and routed livestock into pens by car so that the same animals would be reloaded into the proper cars.

The USDA specified a minimum amount of feed to be provided to animals at rest stops. For cattle, it was 200 pounds of feed or mixed hay and feed per car; for sheep, it was 100 pounds of feed or mixed feed per deck; and for hogs, two bushels of shelled corn per deck. This feed was provided by the stockyard or railroad and billed to the owner.

When animals were reloaded into their cars, the time was noted on the waybill along with the charges for feed. This was also noted on another form that covered the entire train or cut of cars being handled. Cars would generally be cleaned, with old bedding removed and fresh bedding added.

In some cases, livestock would be reloaded in another car. The waybill included a blank box (right of the car initials) for listing the number of the new car in which the livestock was transferred. This was most often done when cattle were off-loaded from foreign-road cars. If a home-road empty car was available, the foreign-road car would be sent

to keep the animals cool. Straw was common in the winter to add warmth. The railroad billed the owner for the service. Owners would sometimes bed the cars themselves to avoid these charges.

Cattle were weighed before loading if the loading facility had a scale; otherwise, cars would be weighed on a track scale as soon as possible after being loaded, and

the weight determined by subtracting the light weight of each car.

Cars were to be loaded by the owners, not the railroad, although railroad personnel could be on hand to make sure cars were loaded properly and to inspect the condition of the livestock to make sure descriptions were accurate as listed on the waybill and contract.

Keeping hogs wet in summer heat was vital in keeping them healthy during transit. This manual twin-handle pump soaks both decks of hogs on Swift Live Stock Express cars in a New York Central train east of Rome, N.Y. *S.K. Bolton Jr.*

home, and the cattle reloaded to the home-road car.

In addition to rest stations, railroads provided trackside watering stands where cars were sprayed down in hot weather. The train would roll by at low speed as nozzles directed water at the sides of stock cars. The photo on this page shows a manual pumping station, and a more modern one is pictured on page 76.

The idea was not to provide drinking water but to keep the bedding soaked, which was especially important for pigs. A common saying in shipping pigs by rail in the summer was "a wet hog is a live hog."

Mainline stock trains

Western railroads operated stock trains in a manner somewhat like passenger trains. Instead of local passenger trains bringing passengers to hubs where they transferred to intercity streamliners, railroads collected loaded stock cars from branches and secondary lines and consolidated them into solid blocks of cars and trains bound for livestock markets such as union stockyards in major cities.

For example, the Chicago & North Western collected stock trains and cuts of stock cars from its western branches and secondary lines at Chadron in western Nebraska. From there, solid trains and blocks of cars were sent 320 miles east to Norfolk, Neb. There, cuts of stock cars were classified and new trains assembled based on final destinations, mainly large union stockyard centers located on the C&NW, which included Omaha, Sioux Falls, Sioux City, South St. Paul, and Chicago.

Because much of the stock traffic was seasonal (heavier in the fall and spring), many stock trains ran as extras but were given priority treatment. Although extras were the lowest-ranked train on a timetable, dispatchers could issue train orders specifying that other trains take sidings for meets. Short of delaying a premiere passenger train (and even that was done sometimes), stock and perishable trains were given priority over most other traffic.

Stock trains could also be run as a first or second section of a regular freight train, especially a perishable train (more on that in Chapter 7). Long cuts of stock cars could often be found at the head end of regular perishable (produce and meat) trains.

Some railroads ran stock trains on a regular basis, giving them their own train number or designation. Among the best known of these was Union Pacific's Daylight Livestock (DLS), train 299, which ran from Salt Lake City to Los Angeles. The railroad inaugurated the service in 1947.

Previous service on this 820-mile trip had taken around 60 hours and required a rest stop at the UP's Las Vegas stockyard en route. When inaugurating the DLS, the

A triple-headed shower wets down the hogs riding in Union Pacific triple-deck cars at Dry Lake, Nev., on a hot July day in 1977. The California Livestock Special is bound for Los Angeles. *Steve Patterson*

railroad scheduled the train for a 27-hour trip, allowing the railroad to close its Las Vegas facility.

To accomplish this, the railroad refurbished a group of 300 stock cars especially for DLS service (see Chapter 2), equipping them with roller bearing trucks and painting them yellow with aluminum roofs and ends, which reflected sunlight in the desert heat. Additional cars were later converted.

The train originated in Ogden at 11:30 a.m. and left Salt Lake City at 12:30 p.m. This departure time allowed it to travel the hottest stretch of desert across southern Nevada at night, with an arrival in Los Angeles at 3:30 p.m. the following day. The train sometimes ran in multiple sections during peak seasons, with train size limited to 85 cars. As a priority train, the DLS rated the railroad's most modern locomotives.

Transported livestock initially included hogs, sheep, and cattle. Hog watering stations (pictured above) were set up at three locations to keep car bedding wet during hot weather.

The livestock was headed for a variety of destinations and customers in the Los Angeles area (more than 25 at first). Through the years, as overall livestock business declined, the major customer became Clougherty Packing Co., the biggest pork packer on the West Coast and maker of Farmer John meats. In 1976, the UP renamed the train the California Livestock Special (CLS).

Hogs came from various Midwest points, including Omaha, Sioux City, and Kansas City, that came together from Council Bluffs to Salt Lake City. By the mid-1970s, the UP had modified many stock cars by adding in-car water and feed stations and retractable side slats. This allowed the railroad to avoid the 36-hour rule and eliminate the Salt Lake City feed and rest station.

The CLS continued until 1985, and after that, the little stock traffic that still existed was carried on regular hotshot intermodal trains. The final run came in 1994, which ended livestock rail service in the United States.

Other stock trains included Santa Fe's train No. 41, the Hog Special, which ran the 869 miles from Belen, N.M., to Los Angeles and the Chicago, Burlington & Quincy's Yellowstone Livestock Express that ran from Laurel, Mont., to Lincoln, Neb.

Stockyard rail operations

As shown in Chapter 4, many union stockyards hosted huge rail operations, with several miles of track within the facility. How a stockyard was served varied by location.

For example, in Chicago, each railroad switched its own trains of cars in and out of the stockyard, with stockyard employees handling all loading and unloading.

Other stockyards were serviced by a separate railroad—which was generally owned by the member railroads that served the stockyard.

One example was Sioux City, served by the Sioux City Terminal Railway, which was owned by the five railroads that served the stockyard: the Chicago & North Western, Chicago, Burlington & Quincy, Great Northern, Illinois Central, and Milwaukee Road.

The SCT not only served the stockyard but also switched all of the surrounding packing companies (except IBP). It operated up to five locomotives at a time, switching nearly 100,000 cars annually in the late 1950s.

Regardless of the operating scheme, inbound loads of livestock were spotted as soon as possible for unloading. Trains arriving in a terminal city had their stock loads pulled immediately upon arrival, with cars delivered directly to the stockyard by a transfer run or delivery to the terminal railroad that serviced the stockyard.

Once cars were unloaded, they were pulled from the platforms and sent to servicing or clean-out tracks. Cars that needed to be reloaded were cleaned and had new bedding added; these cars were then spotted back at platforms as needed.

Home-road cars were loaded for transit on their respective railroads. For example, if 10 cars of cattle at Sioux City were destined for South St. Paul on the Chicago & North Western, C&NW cars were used if available.

Outbound loads were handled in a similar manner. Blocks of cars were pulled from ramps as soon as they were loaded and picked up by or delivered to their respective railroads quickly for placement on priority trains.

Empty stock cars required less urgency, although during peak season, railroads hustled to get empties back quickly for reloading. As cars were unloaded, foreign-road empties were sent toward their home railroads, and home-road empties were reloaded if possible or sent where needed (usually west).

Drovers car

When a carload or more of livestock was shipped, railroads provided free transport for the drovers (cowboys or ranch hands who accompanied the shipment). If this was a small shipment with just one or two drovers, they would generally ride in the caboose with the train crew.

For larger shipments with several drovers (more than a half dozen or so, depending upon the railroad), railroads usually provided a separate drovers car ahead of the caboose.

Drovers cars were generally old cabooses or passenger cars that had been refitted with several bunks. For former coaches or sleeping cars, a stove would be added since there was no steam heat available at the end of a freight train.

Railroad crews did not like sharing space with drovers because drovers tended to smell like the loads they were tending. Drovers were also not bound by Rule G, which required train crews to abstain from alcoholic beverages.

The use and assignment of drovers cars varied by railroad, but most western railroads that handled a lot of livestock traffic operated drovers cars through the 1950s.

Chicago, Burlington & Quincy drovers car 5760 was rebuilt in 1948 from a heavyweight 12-section, one-drawing room sleeper. It's shown here in 1958. *Hol Wagner collection*

Declining service

Stock and meat traffic had begun declining sharply in the 1950s as packers relocated away from metro union stockyards to rural areas near feedlots. Trucks were taking more and more livestock traffic as highways improved and truck size increased.

By the late 1960s, the railroad livestock business had dropped well under 100,000 carloads annually. The few stock cars that remained in service were largely old and obsolete, and railroads couldn't justify rebuilding or replacing these cars. Most railroad-maintained stock pens had been abandoned or torn down.

In the early 1970s, the Santa Fe—long the leading railroad in stock car ownership and stock shipments—and several other western railroads applied to the ICC to discontinue carload rates on western and transcontinental livestock movements. The ICC approved this in 1974, effectively ending most livestock traffic in the country.

A few livestock movements hung on where railroads could ship trainloads or long cuts of cars for specific shippers or vendors. Into the 1980s, the Canadian Pacific and Canadian National both still operated some livestock trains, even though stock service was pretty much gone in the United States.

The last stock shipments in the United States were a remnant of the UP's CLS that brought hogs from a few locations in the Plains states to the Clougherty plant in Los Angeles in cars rebuilt from 60-foot boxcars (see Chapter 2). When the service ended in 1994, the era of shipping animals by rail ended.

Meat traffic

Illinois Central dispatch CC-6 was the IC's premiere freight train on the Iowa Division. It carried solid strings of refrigerator cars loaded with packing house products from several Midwestern cities to eastern markets. Here, the train pauses along the Mississippi River on its journey from Council Bluffs to Chicago in October 1957.
William D. Middleton

Once the packing companies have done their jobs and turned animals into meat, the final products must get to market as soon as possible—a job that was largely the responsibility of railroads through the 1950s. Containing perishable products, refrigerator cars of meat usually traveled in high-priority fast freight trains.

Since meatpacking was centered heavily in various cities of the Midwest, solid trains of refrigerator cars could sometimes be found heading to the major markets in the Northeast. In other areas, strings of reefers wearing the paint schemes of various packing companies could be found in freight trains throughout the country.

An 0-6-0 of the South Omaha Terminal Railway shuttles cars near the Omaha stockyard in the 1940s with a string of Swift refrigerator cars in the background. *Joseph Burkhart*

Coming out of a packing plant, each refrigerator car was destined for a particular branch house, wholesaler, or other customer. This was unlike some produce traffic that was sent on its way without a specific customer. Called *rollers*, these cars were consigned while in transit or sent to yards in large groups for auction or sale to various buyers.

For example, say a day's output at a particular packing plant in Omaha was 40 refrigerator cars. Fifteen cars are destined for branch houses in New York City, 5 for Boston, 5 for Philadelphia, 5 for Los Angeles, and 10 for various small-city branch houses in the Southeast.

The 5 cars bound for Los Angeles would be added to a westbound priority freight. The remaining 35 cars would head eastward, quite likely in a solid block to Chicago. At Chicago, the Indiana Harbor Belt would move the New York, Boston, and Philadelphia cars to their connecting railroad or railroads (possibly together, depending on the routing specified by the shopper). The other 10 cars would be routed toward the Southeast on various routes.

This means that solid trains of meat reefers could be found leaving meatpacking cities, but the farther they traveled, the more the cars were divided and rerouted.

Illinois Central's meat trains

One of the best-known meat-hauling railroads was the Illinois Central, which served several major meatpacking cities in the Midwest including Sioux City, Omaha, Sioux Falls, Waterloo, and Dubuque, as well as Chicago. Through the 1950s and into the 1960s, the IC carried a great deal of this packing-plant traffic on dedicated trains across its Iowa Division—50,000 carloads annually through the late 1950s, more than all other Iowa railroads combined.

The IC's traffic matched the country's overall flow of livestock and meat products, which was mainly west to east. To serve the many packing plants on its line, along with other through perishable traffic, the railroad ran a number of eastbound priority (dispatch to the IC) freight trains dedicated to the business.

Dispatch train CC-6 (Council Bluffs-Chicago, train No. 76) was the hottest train on the division, and possibly the hottest freight train on the entire railroad. Its main purpose was to bring packing-house traffic from the western end of the railroad to Chicago, where connecting railroads would take cars to points throughout the eastern United States.

Train CC-6 was scheduled to leave Council Bluffs, just across the Missouri River from the stockyards and packing plants of South Omaha, every evening around 7 p.m. (Exact train times varied through the years.)

Eastbound packing plant traffic from the Wilson, Armour, Cudahy, and Swift plants (as well as from several smaller packers) would have been pulled in the afternoon by the South Omaha Terminal Railway, which served the city's stockyard and most of its packing plants. The SOT would place outbound cars on appropriate yard tracks for each railroad it served. The Chicago Great Western, Chicago & North Western, Rock Island, and

The Nickel Plate carried a lot of eastbound meat traffic from Chicago and other cities. Here, Berkshire No. 776, with two Swift cars at the head end, rolls a long string of meat reefers near Vermilion, Ohio, in 1957. *H.S. Ludlow*

Milwaukee Road also shared this eastbound traffic from Omaha.

An IC switch job would cross the river and pick up the IC's cars in late afternoon or early evening and bring them back to the IC Council Bluffs yard in time for CC-6's departure. Along with reefers of dressed meat, traffic could include other packing-house products or by-products, stock cars of livestock from the Omaha stockyards bound for eastern markets (including loads of hogs bound for the Rath plant in Waterloo), and other priority freight. The amount of traffic varied by day, with more reefer loads later in the week. Heavy traffic would warrant running the train in multiple sections.

At the same time, similar eastbound trains were leaving Sioux City and Sioux Falls. Illinois Central dispatch SCF-6 (Sioux City-Fort Dodge, train No. 676) left Sioux City at 5:40 p.m. and would have cars from the city's Armour, Cudahy, and Swift plants. It was due in Fort Dodge at 9:35 p.m. Dispatch SFC-6 (train 776), carrying Morrell cars, left Sioux Falls at 4:45 p.m. to make it to Cherokee at 7:45 p.m. If traffic was light, it would be combined there with SCF-6; if traffic was heavy, both trains would run through to Fort Dodge.

Train CC-6 would arrive at Fort Dodge around 10:30 p.m. At Fort Dodge, CC-6 might pick up Sioux City and Sioux Falls meat cars from SCF-6; however, if there were enough loads, SCF-6 would run through to Waterloo as an additional section of CC-6.

The various sections of CC-6 would begin arriving at Waterloo after midnight, and the trains would be classified and blocked based on their connections in Chicago. Reefers would be iced if necessary at the IC's 52-car icing station and platform.

© 2012 Kalmbach Publishing Co.: Rick Johnson

Livestock and meat reefers often traveled together in the same train, as shown here by eastbound Illinois Central CC-6 west of Rockford, Ill., in 1954. *David P. Morgan Library collection*

Rath had a large packing plant in Waterloo (one of the largest pork plants in the country) and sometimes delivered more than 100 cars a day to the IC. Much of this traffic would have already headed east on Dispatch WC-2 (train No. 78), a Waterloo-Chicago train due out of Waterloo at 8 p.m. Late Rath traffic could be added to CC-6.

The various sections of CC-6 headed out of Waterloo and proceeded eastward toward Chicago. The goal was to make a connection with the Indiana Harbor Belt at Broadview (just west of Chicago) by 1:30 in the afternoon, which allowed the IHB to ice the cars and forward the meat loads to the IHB's connecting eastern railroads for departure on their respective priority freights later that day.

Another priority train on the Iowa Division was AC-2 (Albert Lea-Chicago). Nicknamed *The Apple*, AC-2 carried mainly fruit traffic from the Northwest picked up at interchange in Albert Lea. The train could also carry meat traffic from nearby Austin and from South St. Paul packing companies via the Minneapolis &

The South Omaha Terminal assembled cuts of cars for several railroads, including (from left) the Chicago & North Western, Rock Island, and Illinois Central. The Rock and IC transfers are ready to roll on this October evening in 1957. *William D. Middleton*

St. Louis. Additional meat loads could be added at Waterloo en route to Chicago.

In addition, Council Bluffs-Chicago train CC-4 could also carry some meat traffic. Train CC-4, due out of Council Bluffs in early morning, primarily carried produce

from the West from the Union Pacific. These trains (other than CC-6) could also pause to pick up loaded meat reefers at Dubuque.

Other meat haulers

The Chicago Great Western and Milwaukee Road also carried

The Erie's icing platform at Marion, Ohio, handled a lot of meat and other perishable traffic. Here, carts on the upper-level platform service a pair of Wilson meat reefers. The two Pacific Fruit Express reefers at right will receive chunk ice from the main platform. A large ice plant is in the background. *Erie Railroad*

Crushed ice is dumped from a cart on the upper level through a funnel that rides on tracks at the platform edge. *Erie Railroad*

Midwestern meat traffic in a similar manner to the Illinois Central, combining cars from several points and getting them to Chicago.

Several railroads carried a lot of eastbound meat traffic out of Chicago, including Nickel Plate,

New York Central, Pennsylvania, and Erie. This included cars from Chicago packing plants as well as those passing through the city from other Midwestern plants.

A well-known meat train was Erie No. 98, running from Marion, Ohio, to New York City. The

train always carried a great deal of Chicago packing-house traffic as well as produce and other perishables from the West and Southwest. The train left Marion at 9 a.m. and made a regular stop to ice cars at Hornell in western New York around midnight. The ice station there had a double-deck platform to ice both bunker and brine-tank cars. Number 98 continued on to Jersey City, arriving in late afternoon, from where cars were distributed in New York City.

Other notable meat runs were Nickel Plate Road trains MB-18 from East St. Louis to Buffalo and CB-12 from Chicago to Buffalo.

Meat reefer handling

Refrigerator cars of packing house products (and other perishable cars) were given priority handling when arriving at terminals. They often traveled at the head end of trains to make it easy to switch them out. They were often the first cars pulled by switchers when arriving at a terminal. Final delivery could be performed by dedicated switching jobs in larger cities or by local freights for

A worker shovels salt into a reefer via a rolling chute at the Chicago, Burlington & Quincy's icing dock in Denver in 1949. The conveyor chain on the platform at right carries block ice along the platform. *Earl Cochran*

destinations outside of large metro areas.

Once empty, meat reefers were returned directly to their owners or lessors. As private-owner cars, they would not be sent to another company's packing plant or loaded for their return trip.

Cars returning empty were not given the same priority as loaded cars—they usually traveled in general merchandise freight trains. However, packing companies still wanted cars returned quickly, so they could turn them around for their next loads. Packing companies specified return points for each of their cars (generally the same routing as outbound shipments). Back at the packing plant, cars would be spotted at a plant's clean-out tracks or yard. Empty cars might also be kept at a nearby railroad yard or siding if the packer didn't have room for all returning empties.

Unlike produce cars, meat cars stayed with their owners. Although you might see a train with a combination of Armour, Swift, and Morrell cars, an Armour car would never be spotted at a Swift packing plant or branch house, and a Swift car wouldn't show up in a clean-out track at a Morrell plant.

Icing stations

Although a few mechanical refrigerator cars had been placed in meat service by the 1960s, ice reefers dominated until the end of meat reefer service. A key logistical challenge in handling perishable traffic was replenishing ice in the cars' bunkers and brine tanks.

Icing—the process of adding ice to cars—was done at icing platforms or icing stations. Most medium- to large-size packing plants had their own icing stations, or the service was provided by the terminal railroad serving the plant and stockyard area. Railroads operated icing stations along their main perishable-traffic routes as well as at major yards. The largest of these, with tracks on either side

of a long platform, could service an entire train at once.

The specific designs of platforms differed by railroad and era, but most looked similar. A tall deck (car roof height) on an open frame ran parallel to a track (or tracks might be located on both sides of the platform). Platform length varied by installation, but some that served a lot of perishable traffic were equipped to service 50 or more cars at once. Many other locations had smaller icing stations (5–20 cars long). At these, an inbound train would have to move one or more times during the icing process to serve all the cars.

An ice house would be located adjacent to the platform. Large blocks of ice, 300 pounds or heavier, would be pulled from the ice house to the platform and along it by a series of chains on the deck.

For most of the 1900s, the majority of ice was made mechanically by refrigeration. If it wasn't made at the ice house itself, it

The night icing crew at the Illinois Central's Waterloo, Iowa, platform adds crushed ice to an Armour refrigerator car in 1957. *William D. Middleton*

would be brought in from another facility by railcars—generally older reefers stenciled for company ice service only. Prior to the widespread use of mechanically made ice around 1900, many railroads and ice companies cut ice from frozen lakes in the winter, then moved the blocks to heavily insulated ice houses where they were stored until needed.

As Chapter 3 explained, around 60 percent of meat reefers used brine tanks, which required a mix of crushed ice and salt that was poured into the roof hatches. The remainder of meat reefers, along with almost all produce cars, had standard ice bunkers, which used chunk ice chopped from large blocks.

To serve both types of cars, many platforms had an additional deck above the main (car-height) deck. To ice a standard bunker car, hinged walkways were dropped from the main deck to the car roof. One worker would slide a block of ice onto the car,

where a second worker would position it over the open hatch and then chop it into chunks that fit into the opening.

For a brine-tank car, ice was crushed in the plant and brought on the platform in carts. The ice would be dumped into the tank through the roof hatch via a spout or cart. Another cart loaded with salt would dump its load atop the ice.

Major icing facilities began mechanizing the process in the 1950s, usually by the use of a large machine that traveled on rails along the platform. The machine would take ice from the conveyor chains on the platform and crush it to the size needed for each car. Arms extending over the tracks delivered the ice into the hatches.

Ice trucks were sometimes used to ice individual cars or small blocks of cars where platforms weren't available. This could be done by a local ice dealer or the shipper.

An older 36-foot meat reefer required an initial load of 6,000–7,000 pounds of ice, while many larger (40- and 42-foot) cars could hold 9,000–10,000 pounds of ice. The amount of ice needed depended upon the size of the cars, the load, and the weather—the hotter the weather, the more ice that was needed and the more frequently it would have to be added. A top-off of 2,000–3,000 pounds was typical, with re-icing at least once per day until a car reached its destination.

Icing a train

Solid trains of perishables—whether meat, produce, or a combination—would often be iced at once. If a platform could handle an entire train, the train would pull up next to the dock. In advance of the train's arrival, ice would have been distributed along the length of the platform and crews would be waiting.

Once the train was stopped, workers began by dropping the walkways from the platform to the car roofs. Workers would open the hatches and plugs, and a worker would walk along the cars and estimate how much ice would be required for each bunker or brine tank and chalk the amount on the hatches. Crews would then add the ice.

An efficient crew could ice a car in a few minutes. How long it took to ice an entire train depended upon how many crews were working. A 40-car train with five crews working could be iced in 20–30 minutes.

To ice a train or long block of reefers if the platform was shorter, a train would have to move one or more times. This would obviously increase the time required for icing.

In many cases, only a block of cars, or a few select cars, on a particular train would require icing. Again, train crews and icing crews would know which cars they were. Because of this, and to

A pair of Morrell reefers are tucked in front of the caboose on this hotshot eastbound Baltimore & Ohio freight leaving Willard, Ohio, in 1965. *J.W. Barnard Jr.*

make it easier to switch cars out quickly at terminals, reefers were often kept together in a block (and at the head of a train).

Every reefer wouldn't require replenishment at every icing station since cars were often coming from multiple sources, going to multiple destinations, and carrying a variety of perishable products.

Consider the IC's Iowa Division. The railroad had an icing station at Waterloo, but not all reefers stopped there. Most meat cars on CC-6 had been loaded around 10 hours earlier and could go through to Chicago before being re-icing. However, all cars of a produce train arriving from the west, depending on where it originated, might require re-icing.

By the end of the 1960s, ice refrigerator service was dwindling. Pacific Fruit Express took its icing platforms out of service in 1973; the Santa Fe had already done so in 1971, and most other railroads followed suit. By then, the era of the ice-bunker meat refrigerator car was done.

Modeling meat traffic

Modeling meat trains and operations provides some fascinating potential, even if your layout

The final destination for most meat reefers is to be set out at a local branch house or food wholesaler. Here, a North American reefer leased to Greenlee Packing is switched in Philadelphia in 1964. *J. Michael Gruber collection*

doesn't include a packing plant or branch house. If your layout includes a division-point yard, you can model icing platform operations. You can model inbound trains with meat cars that are combined to form another train or model trains that are combined or run in multiple sections.

If you model smaller cities, you can include a packing company branch house and the meat reefers that are delivered to it regularly.

For the ultimate in meat and stock operations, you could model a large city terminal railroad, including a union stockyard and multiple packing companies. Switch crews would be kept busy pulling loads of cars from packers in the afternoon, assembling outbound cuts of cars in the evening, and delivering empty cars to the packing plants overnight for loading the following day.

However you choose to do it, modeling meatpacking and livestock operations will be a colorful addition to any layout set in the 1960s or earlier.

Bibliography

Books

The American Railroad Freight Car, John H. White Jr., Johns Hopkins University Press, 1993

America's Historic Stockyards: Livestock Hotels, J'Nell L. Pate, Texas Christian University Press, 2005

Beyond Beef: The Rise and Fall of the American Cattle Culture, Jeremy Rifken, Dutton, 1992

Billboard Refrigerator Cars, Richard H. Hendrickson and Edward S. Kaminski, Signature Press, 2008

By-products in the Packing Industry, Rudolf A. Clemen, University of Chicago Press, 1927

Chicago Union Stock Yards, Union Stock Yard and Transit Co. of Chicago, 1953

Classic Freight Cars, Volume 3: 40-Foot Refrigerator Cars, John Henderson, H&M Productions, 1993

The Great Yellow Fleet, John H. White, Golden West Books, 1986

Pacific Fruit Express, Anthony W. Thompson et al., Signature Press, 1992

The Postwar Freight Car Fleet, Larry Kline and Ted Culotta, National Model Railroad Association, 2006

Railroad Facilities for Handling Livestock at Shipping Points in the Corn Belt Region, Knute Bjorka, U.S. Bureau of Agricultural Economics, 1943

Refrigerator Car Color Guide, Gene Green, Morning Sun Books, 2005

The Station Agent's Blue Book, O.B. Kirkpatrick, Kilpatrick Publishing Co., 1937

Stock Car Cyclopedia, Vol. 1, Robert L. Hundman, Hundman Publishing, 2007

Stock Cars of the Santa Fe Railway, Frank M. Ellington et. al., Railroad Car Press, 1981

The Story of Meat, Robert B. Hinman, Swift & Co., 1933

Territories of Profit, Gary Fields, Stanford University Press, 2004

Visitor's Bulletin, Swift & Co., 1933

Periodicals

Car Builder's Cyclopedia, Simmons-Boardman, various issues

The Freight Traffic Red Book, various issues

The Official Railway Equipment Register, various issues

Railway Age, various issues

"40-foot Mather Stock Cars," *RailModel Journal*, February 1997

"Armour Car Lines Steel Refrigerator Cars," Ed Hawkins, *Railway Prototype Cyclopedia*, No. 21, 2010

"C&NW Stock Car, 1950s Rebuilds," Jeffrey M. Koeller, *Mainline Modeler*, November 2003

"Everything but the Squeal: The Milwaukee Stockyards and Meat Packing Industry, 1840-1930," Paul E. Geib, *Wisconsin Magazine of History*, Autumn 1994

"General American 37-Foot Meat Reefers, Patrick C. Wider, *Railway Prototype Cyclopedia*, No. 14, 2006

"Harriman Standard 36-foot Stock Cars," Richard Hendrickson, *RailModel Journal*, October 2006

"Iowa Division is King Size in Every Way," *Illinois Central Magazine*, September 1951

"Meat Reefers," Martin Lofton, *Mainline Modeler*, 3 parts: February, March, April 1992

"Meat Train," William D. Middleton, *Trains*, October 1958

"Modeling the Meat-Packing Industry," Jeff Wilson, *Model Railroader*, 2 parts: January and February 2010

"NP's Big Pig Palace Cars," Chuck Yungkurth, *Model Railroader*, April 1982

"Reefers of the Union Refrigerator Transit Co.," Al Westerfield, *RailModel Journal*, July 1992

"Reefers, Stock Cars and Tank Cars of the Swift Fleet," Martin Lofton, *RailModel Journal*, February 1993

"Rolling Livestock," Jeff Wilson, *How to Build Model Railroads* (special issue of *Model Railroader*), 2007

"SL-SF Stock Car," Robert L. Hundman, *Mainline Modeler*, November 2001

"Stock Cars: Their Development and Use," John Nehrich, *Mainline Modeler*, three parts: June, July, August 1990

"Trackside Stock Pens for Your Layout," B. Paul Chicoine, *Model Railroader*, April 1993

Historical society publications

"Burlington Billboard Reefers," Hol Wagner, *Burlington Bulletin*, No. 28, Burlington Route Historical Society

"The Chicago Great Western and the Thursday Night Meat Train," Dick Wilson, *North Western Lines*, 2006 No. 3, Chicago & North Western Historical Society

"Daylight Livestock," Terry Metcalfe, *Streamliner*, Vol. 1, No. 1, Union Pacific Historical Society

"Hauling Stock on the NP," Rufus Cone, *The Mainstreeter*, Spring 1987, Northern Pacific Railway Historical Society

"Illinois Central's Meat Trains on the Iowa Division," *Central Standard Times*, Issue 95-5, October 1997, Chicago Central Historical Society

"Livestock Traffic," Hol Wagner, *Burlington Bulletin*, No. 25, Burlington Route Historical Society

"Rath Packing," Paul Michelson, *Green Diamond*, August 1998, Illinois Central Historical Society

"Reefers," Rod Masterson et al, Burlington Bulletin, No. 12, Burlington Route Historical Society

"Santa Fe's Livestock Service, Part 1: History and Operations," Matt Zebrowski, *The Warbonnet*, Third Quarter 2001, Santa Fe Railway Historical & Modeling Society

"Santa Fe's Livestock Service, Part 2: The Stock Cars," Richard H. Hendrickson, *The Warbonnet*, Fourth Quarter 2001, Santa Fe Railway Historical & Modeling Society

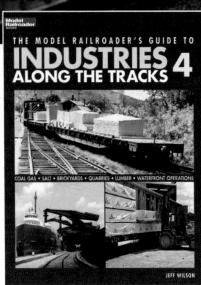
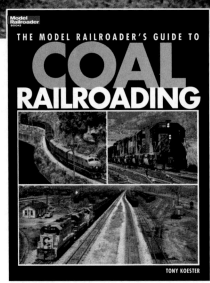
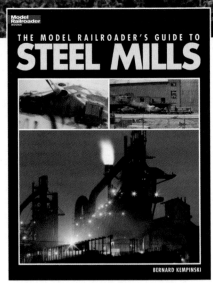